THE 2011 NATIONAL ELECTRICAL CODE BOOK OF IN-DEPTH CALCULATIONS

VOLUME 4

Calculations Worksheets

Alvin J. Walker

THE 2011 NATIONAL ELECTRICAL CODE BOOK
OF IN-DEPTH CALCULATIONS

VOLUME 4

© 2015 Alvin J. Walker

Portions of this material are reproduced from NFPA70®- 2011, *National Electrical Code*®,
Copyright © 2010, National Fire Protection Association, Quincy, MA.
This reprinted material is not the complete and official position of the NFPA on the referenced
subject, which is represented only by the standard in its entirety. The calculation accuracy
of any equation is the sole responsibility of the author and not the NFPA.

NFPA70®, *National Electrical Code* and *NEC*® are registered trademarks of the
National Fire Protection Association, Quincy, MA.

ISBN 13: 978-0-9831358-5-2

LCCN 2015907070
First Edition
1 2 3 4 5 6 7 8 9 10

Walker & Walker Electrical Consultants
For more information, please contact
Alvin Walker
318-393-6841
www.alvinwalker.com

TABLE OF CONTENTS

INTRODUCTION TO VOLUME 4

The following worksheets are provided for the sole and intended purpose of familiarizing the user with the necessary guidelines and procedures for performing relative load calculations. Each worksheet should be constantly and thoroughly reviewed to gain knowledge of its use and how each specific application of the National Electrical Code (NEC) is applied.

As a whole, these worksheets provide detailed instructions for calculating both standard and optional load calculations for single and multifamily dwellings along with other worksheets that are second-to-none for performing commercial and industrial load calculations and affiliated utilization equipment.

Because of the possibility of one either not understanding or misinterpreting the provisions of Sections 220.54 *and* 220.55 (Household Electric Clothes Dryers *and* Household Range and Other Cooking Appliances) individual worksheets are included.

Again, these worksheets are an honest and best-effort representation of the author's desire to assist the users with exclusive NEC requirements where electrical calculations are necessitated.

STANDARD LOAD CALCULATIONS
FOR ONE-FAMILY DWELLING
<NEC Reference Article 220 - Parts II and III >

---

1. GENERAL LIGHTING and RECEPTACLE LOADS <NEC References - 220.12, Table 220.12, 220.14(J), 220.42 and Table 220.42> (Open porches, garages, unused or unfinished spaces not adaptable for future use not included.)

_____SF + _____SF x 3VA = _____VA
(Dwelling's outside dimensions)   (unfinished space adaptable for future use)   **(1)**

2. SMALL APPLIANCES CIRCUIT LOAD (Portable) <NEC References - 220.52(A) and 210.11(C)(1)> (At least two (2) small appliance branch circuits must be included.)

1500 VA x _____ = _____ VA
                (No less than 2 circuits)   **(2)**

3. LAUNDRY CIRCUIT LOAD <NEC References - 220.52(B) and 210.11(C)(2)> (At least one laundry circuit must be included.)

1500 VA x _____ = _____ VA
                (No less than 1 circuit)   **(3)**

TOTAL [Lines **(1)** - **(3)**] = _____ VA
(If Total VA is less than or equal to 120,000VA, step **c.** is not required)

APPLY DEMAND FACTORS <NEC References - 220.42 and Table 220.42>

a. First 3000VA or less of above TOTAL (at 100 percent) = _____VA

b. _____ x .35   = _____VA
        (Total VA - 3001VA up to 117,000VA)

c. _____ x .25   = _____VA
        (Remainder of TOTAL VA exceeding 120,000VA)*

    *(_____ VA (TOTAL) – 120,000 VA = _____ VA)

TOTAL (Lines **a.** - **c.**)   = _____ VA
(Derated General Lighting and Receptacle Loads,   (Enter value below - LINE
 Small Appliance and Laundry Circuit Loads)   and NEUTRAL)

GENERAL LIGHTING and RECEPTACLE, SMALL APPLIANCE and LAUNDRY LOADS

1. - 3. LINE LOAD   NEUTRAL LOAD

_____ VA   _____ VA

4. APPLIANCE LOADS (Fastened-In-Place) <NEC Reference - 220.53> (Use nameplate rating of each appliance. Electric ranges, dryers, space-heating equipment or air-conditioning equipment not included.)

120V Appliances	VA Rating
1.	
2.	
3.	
4.	
5.	
6.	
7.	
8.	
9.	
10.	
	_____ (Total 120V Appliances)

208V or 240V Appliances	VA Rating
1.	
2.	
3.	
4.	
5.	
6.	
7.	
8.	
9.	
10.	
	_____ (Total 240V Appliances)

APPLIANCES TOTAL (120V and 208V or 240V) = _____ VA

APPLY DEMAND FACTOR (Applicable, when number of above appliances exceeds four (4) or more.)

Appliances Total VA x .75 = _____ VA

APPLIANCE LOADS

4. LINE LOAD   NEUTRAL LOAD (Refer to Conditions)

_____ VA   _____ VA

Conditions
When appliance loads are line-to-neutral (common) [240/120V (208/120V)] or neutral (120V) connections, NEUTRAL LOAD is same as LINE LOAD.

-or-

When appliances are line-to-line (208V or 240V) loads NEUTRAL LOAD equals,

$$\underline{\hspace{2cm}} - \underline{\hspace{2cm}} \times .75^+ = \underline{\hspace{2cm}}$$
(Line Load)   (Line-to-Line Loads)   (NEUTRAL LOAD)

$^+$Demand Factor only applied when difference still results to 4 or more appliances.

5. CLOTHES DRYER <NEC References - 220.54 and Table 220.54> (Use 5000W (VA) or nameplate rating, whichever is larger). Refer to, **(Worksheet E** [Volume 4]) Demand Load Calculations for Household Electric Dryers. *Applies only when dryers are supplied by 240/120V.

CLOTHES DRYER

5.   LINE LOAD   NEUTRAL LOAD
                                           <NEC Reference - 220.61(B)(1)* or (C)(1)>

        _____ VA   _____ VA
                                           *(70 percent of line load)

6. ELECTRIC RANGE or OTHER COOKING APPLIANCES <NEC References - 220.55 and Table 220.55> Refer to, **(Worksheet F** [Volume 4]) Demand Load Calculations for Household Electric Ranges and Other Cooking Appliances. Line-to-line electric range or other cooking appliance loads must not be included in neutral load calculations. *Applies only when electric range or other cooking appliances are supplied by 240/120V.

RANGE/COOKING APPLIANCES

6.   LINE LOAD   NEUTRAL LOAD
                                           <NEC Reference - 220.61(B)(1)* or (C)(1)>

        _____ VA   _____ VA
                                           *(70 percent of line load)

7. HEATING and AIR-CONDITIONING (AC) EQUIPMENT <NEC References - 220.50, 220.51*, 220.60, Table 430. 248 and 440.6(A)> (Include VA rating of air handler [blower]). If heat pump, add compressor and maximum amount of electric heat that will be energized simultaneous [at the same time]. Apply, if Heat Pump is used entirely for heating and air-conditioning load. If not, use larger of Heating and Air-Conditioning loads. Larger load to include all loads that could operate at the same time, *example*: where Heat Pump is used for supplemental needs add to larger load if applicable. Use the following formulas to perform required load calculations.

Worksheet A — Standard Load Calculations for One-Family Dwelling

Heating Load

(1) Electric Heat - _____ W (VA) x _____ x _____ = _____ VA
         (Unit Rating)$^+$ (Percentage)*   (No.)***   (Total)

(2) Electric Heat - _____ W (VA) x _____ x _____ = _____ VA
         (Unit Rating)$^+$ (Percentage)*   (No.)***   (Total)

(3) Electric Heat - _____ W (VA) x _____ x _____ = _____ VA
         (Unit Rating)$^+$ (Percentage)*   (No.)***   (Total)

$^+$If calculation required, use the following formula.   TOTAL HEAT = _____ VA
  (V x A) = Unit Rating   Combined)

AC Load

(1) AC - _____ VA^a + _____ VA^b x _____ = _____ VA
      (Voltage x amps**)   (Voltage x amps**)   (No.)***   (Total)

(2) AC - _____ VA^a + _____ VA^b x _____ = _____ VA
      (Voltage x amps**)   (Voltage x amps**)   (No.)***   (Total)

(3) AC - _____ VA^a + _____ VA^b x _____ = _____ VA
      (Voltage x amps**)   (Voltage x amps**)   (No.)***   (Total)

                             TOTAL AC = _____ VA
                                            (Combined)

Heat Pump Load

(1) Heat Pump - _____ VA^a + _____ VA^b +
            (Voltage x amps**)   (Voltage x amps**)
           _____ W(VA) x _____ = _____ VA
           (Maximum Electric Heat)   (No.)***   (Total)

(2) Heat Pump - _____ VA^a + _____ VA^b +
            (Voltage x amps**)   (Voltage x amps**)
           _____ W(VA) x _____ = _____ VA
           (Maximum Electric Heat)   (No.)***   (Total)

(3) Heat Pump - _____ VA^a + _____ VA^b +
            (Voltage x amps**)   (Voltage x amps**)
           _____ W(VA) x _____ = _____ VA
           (Maximum Electric Heat)   (No.)***   (Total)

                    TOTAL HEAT PUMP = _____ VA
                                         (Combined)

^{a}Compressor ^{b}Fan Motor
**RLA (running load amps) or FLC (full-load current) ***Number of units with identical operating characteristics.

LINE LOAD = _____ VA + _____ VA + _____VA
$\qquad\qquad$ (Larger Load) $\qquad$ (Other[s])• $\qquad$ (Total)
•Blowers/Air Handlers, heat pump(s) for supplement needs, etc. – if applicable. (Use [V x A] when VA not given).

HEATING and AC

7. $\qquad$ LINE LOAD $\qquad\qquad\qquad$ NEUTRAL LOAD (120V motors only)

$\qquad$ _____ VA $\qquad\qquad\qquad$ _____ VA

8. LARGEST MOTOR <NEC References - 430.17*, 430.24, 440.7* and 440.33> (Use motor with highest current per NEC 430.17 and 440.7)

a. _____
$\qquad$ *([Identify] Largest Motor based on highest FLC)

b. _____ VA x .25 = _____VA
$\qquad$ (Largest Motor)

LARGEST MOTOR

8. $\qquad$ LINE LOAD $\qquad\qquad\qquad$ NEUTRAL LOAD (120V motors only)

$\qquad$ _____ VA $\qquad\qquad\qquad$ _____ VA

TOTAL DEMAND LOAD (LINE and NEUTRAL) (List each computed line and neutral loads below and total lines 1. - 8.)

	LINE LOAD	NEUTRAL LOAD
1. - 3. General Lighting and Receptacles Loads, Small Appliances and Laundry Circuit Loads	_____	_____
4. Appliances	_____	_____
5. Clothes Dryer	_____	_____
6. Electric Range or Other Cooking Appliances	_____	_____
7. Heating and Air-Conditioning (AC) Equipment	_____	_____
8. Largest Motor	_____	_____
TOTAL DEMAND LOAD (VA) =	_____	_____

DWELLING'S OPERATING LINE VOLTAGE - _____ V
(Given operating voltage or as determined per test examination)

CALCULATE MINIMUM LINE and NEUTRAL LOADS
(Divide Total Demand Load [**VA**] by operating line voltage [**V**])

LINE LOAD = _____ **VA** / _____ **V** = _____ **A**

NEUTRAL LOAD* = _____ **VA** / _____ **V** = _____ **A**

*Where the feeder or service neutral load exceeds 200A, NEC 220.61(B)(2) permits the load to be reduced by 70 percent. However, this reduction is not permitted if the feeder or service neutral load is supplied from any portion of a 3-wire circuit consisting of 2 ungrounded conductors (1ϕ-208/120V) and the neutral of a 4-wire, 3-phase, wye-connected system (3ϕ-208/120V). Complete the following to determine the permitted Neutral Load.

(1) _____ **A** − 200A = _____ x .70 = _____ **A**
       (Neutral Load)   (Remainder)   (Permitted Reduction)

(2) _____ **A** + 200A = _____ **A**
     (Permitted Reduction)   (Permitted Neutral Load)

SIZE SERVICE (Size of service based on calculated LINE LOAD)

SIZE SERVICE REQUIRED (minimum) _____ **A**

SIZING FEEDER/SERVICE CONDUCTORS - (Based on the calculated LINE LOAD and NEC References.) *For* 240/120V, 3-Wire, Single-Phase Dwelling Services and Feeders *up to 400 amperes* only. <NEC References - 310.15(B)(7) and Table 310.15(B)(7)> *For* other [3-Wire, Single-Phase] or [Three-Phase] Services and Feeders. <NEC References - 215.2(A), 230.42(A), 240.4(B) & (C), 310.10(H), 310.15(B)(2) & (3) and Table 310.15(B)(16)>

FEEDER/SERVICE CONDUCTORS _____

SIZING NEUTRAL CONDUCTOR - (Based on the calculated NEUTRAL LOAD and NEC References.) <NEC References - 215.2(A)(2), 220.61, 230.42(C), 250.24(C), 310.10(H), 310.15(B)(2), (3) & (5), 310.15(B)(7) and Table 310.15(B)(16)>

NEUTRAL CONDUCTOR(S) _____

SIZING GROUNDING ELECTRODE CONDUCTOR <NEC References - 250.24(C), 250.66 & Table 250.66 and Table 8 of Chapter 9>

GROUNDING ELECTRODE CONDUCTOR _____

OPTIONAL LOAD CALCULATIONS
FOR ONE-FAMILY DWELLING
\<NEC Reference 220.82 - Part IV\>

The Optional Load Calculation for a **One-Family Dwelling** is only permitted for use (instead of the method [standard] outlined in Part III of Article 220) when a dwelling unit is served by a **single-phase, 3-wire, 240/120-volt** or **208Y/120-volt** service or feeder and where the service or feeder conductors supplying the dwelling unit are rated for **100 amperes** or more. Because this optional method does not include provisions for calculating the neutral load the standard method must be applied in accordance with NEC 220.61.

NEC 220.82(B) - General Loads (1. - 4.)

1. **GENERAL LIGHTING and RECEPTACLE LOADS** \<NEC Reference - 220.82(B)(1)\> (Open porches, garages, unused or unfinished spaces not adaptable for future use not included.)

_____SF + _____SF x 3VA = _____ VA
(Dwelling's outside dimensions)   (unfinished space adaptable for future use)   **(1)**

2. **SMALL APPLIANCES and LAUNDRY CIRCUIT LOAD** \<NEC Reference - 220.82(B)(2)\> (At least two (2) small appliance and one (1) laundry branch circuit must be included.)

1500 VA x _____ = _____ VA
(No less than 3 circuits)   **(2)**

3. **APPLIANCES and MOTORS LOADS** \<NEC References - 220.82(B)(3) and (4)\> (Use the nameplate rating of all appliances that are fastened in place, permanently connected, or located to be on a specific circuit. This includes: ranges, wall-mounted ovens, counter-mounted cooking units, clothes dryers that are not connected to the laundry branch circuit as required (per item **2.**), water heaters *and* the nameplate ampere or kVA rating of all permanently connected motors not included in this item **(3)**. **NOTE:** Heating and Air-conditioning equipment is not included.)

List all appliances and motors with VA rating. Where name plate rating is listed in amperes, use either formulas: V(volts) x A(amperes) = **VA** *or* **VA** = kVA / 1000.

Appliances and Motors	VA rating
Cooktop	_____
Dishwasher	_____
Disposal	_____
Dryer	_____
Freezer	_____
Microwave oven	_____

Range _____
Refrigerator _____
Trash compactor _____
Wall-mounted oven _____
Water Heater _____

_____ _____
_____ _____
_____ _____
_____ _____
_____ _____
_____ _____
_____ _____
_____ _____

Total Appliances VA rating = _____
                              (3)

4. **APPLY DEMAND FORMULA** <NEC Reference - 220.82(B)>

$$\text{TOTAL} = \underline{\hspace{3cm}} \text{VA}$$
[Add loads **(1)** - **(3)**]

   a. First 10kVA (10,000VA) or less of above TOTAL = _____ VA
   (At 100 percent)

   b. _____ x .40   = _____ VA
   (Remainder of TOTAL exceeding 10,000VA)*

       *(_____ VA – 10,000VA = _____ VA)

   c. Total (add lines **a.** and **b.**)   = _____ VA

5. **HEATING AND AIR-CONDITIONING LOAD** <NEC Reference - 220.82(C)> (Use the largest load of selections **a. - f.** as follows.)

   a. **100 percent** of the nameplate rating(s) of the air-conditioning and cooling.

$$\text{AC} = \underline{\hspace{2cm}} \text{VA}$$
(TOTAL)

   b. **100 percent** of the nameplate rating(s) of heat pump when the heat pump is used without any supplemental electric heating.

$$\text{HEAT PUMP} = \underline{\hspace{2cm}} \text{VA}$$
(TOTAL)

c. 100 percent of the nameplate rating(s) of the heat pump compressor and **65 percent** of the supplemental electric heating for central electric space-heating systems.

HEAT PUMP _____ VA + (SUPPLEMENTAL HEAT x .65) = _____ VA
(TOTAL)

If the heat pump compressor is prevented from operating at the same time as the supplementary heat, it does not need to be added to the supplementary heat for the total central space heating load.

SUPPLEMENTAL HEAT x .65 = _____ VA
(TOTAL)

d. 65 percent of the nameplate rating(s) of electric space heating if less than four (4) separately controlled units.

ELECTRIC HEAT x .65 = _____ VA
(TOTAL)

e. 40 percent of the nameplate rating(s) of electric space heating if four (4) or more separately controlled units.

ELECTRIC HEAT x .40 = _____ VA
(TOTAL)

f. 100 percent of the nameplate ratings of electric thermal storage and other heating system where the usual load is expected to be continuous at the full nameplate value. Systems qualifying under selection **f.** shall not be calculated under any other selection in 220.82(C).

ELECTRIC THERMAL STORAGE and OTHER = _____VA
(TOTAL)

TOTAL DEMAND LOAD
(Add lines **4c.** and **5.** [the largest selection])

LINE LOAD

4c. General Lighting and Receptacle, Small Appliances and Laundry Circuit Loads and Appliances = _____ **VA**

5. Heating and Air-Conditioning (AC) Equipment = _____ **VA**

TOTAL DEMAND LOAD = _____ **VA**

DWELLING'S OPERATING LINE VOLTAGE - _____ **V**
(Given operating voltage or as determined per test examination)

CALCULATE MINIMUM LINE LOAD
(Divide Total Demand Load **[VA]** by operating line voltage **[V]**)

LINE LOAD = _____ **VA** / _____ **V** = _____ **A**

SIZING FEEDER/SERVICE CONDUCTORS - (Based on the calculated LINE LOAD and NEC References.) *For* 240/120V, 3-Wire, Single-Phase Dwelling Services and Feeders *up to 400 amperes* only. <NEC References - 220.82, 310.15(B)(2) & (3), 310.15(B)(7) and Table 310.15(B)(7)> *For* other 3-Wire, Single-Phase Services and Feeders. <NEC References - 215.2(A), 220.82(A), 230.42(A), 240.4(B) & (C), 310.10(H), 310.15(B)(2) & (3) and Table 310.15(B)(16)> Ampacity of Service or Feeder conductors must be **100A** or greater.

FEEDER/SERVICE CONDUCTORS _____

SIZING NEUTRAL CONDUCTOR - There is no optional load calculation procedure for sizing the neutral conductor. To size the neutral conductor use standard load calculation for one-family dwelling or size the neutral conductor the same as feeder/service [ungrounded] conductors. <NEC References - 215.2(A)(2), 220.61, 220.82(A), 230.42(C), 250.24(C), 310.10(H), 310.15(B)(2), (3) & (5), 310.15(B)(7) and Table 310.15(B)(16)> As an additional reference, refer to (**Worksheet A** - Volume 4) STANDARD LOAD CALCULATIONS FOR ONE-FAMILY DWELLING.

NEUTRAL CONDUCTOR _____

SIZING GROUNDING ELECTRODE CONDUCTOR <NEC References - 250.24(C), 250.66 & Table 250.66 and Table 8 of Chapter 9>

GROUNDING ELECTRODE CONDUCTOR _____

STANDARD LOAD CALCULATIONS
FOR MULTIFAMILY DWELLING
<NEC Reference Article 220 - Parts II and III >

---

1. GENERAL LIGHTING and RECEPTACLE LOADS <NEC References - 220.12, Table 220.12, 220.14(J), 220.42 and Table 220.42> (Open porches, garages, unused or unfinished spaces not adaptable for future use not included.)

Floor Plan 1 _____ SF + _____ SF x _____ 3VA = _____ VA
              (Dwelling's outside dimensions)   (future use)   (No. of units)

Floor Plan 2 _____ SF + _____ SF x _____ 3VA = _____ VA
              (Dwelling's outside dimensions)   (future use)   (No. of units)

Use same formula for additional floor plans and include in Total.

Total (Floor Plans) = _____ VA
(1)

2. SMALL APPLIANCES CIRCUIT LOAD (Portable) <NEC References - 220.52(A) and 210.11(C)(1)> (At least two (2) small appliance branch circuits must be included.)

1500 VA x _____ x _____ = _____ VA
          (No less than 2 circuits)   (number of units)   **(2)**

3. LAUNDRY CIRCUIT LOAD <NEC References 220.52(B) and 210.11(C)(2)> (At least one laundry circuit must be included)

1500 VA x _____ x _____ = _____ VA
          (No less than 1 circuit)   (number of units)   **(3)**

TOTAL [Lines **(1)** - **(3)**] = _____ VA
(If Total VA is less than or equal to 120,000VA, step **c.** is not required)

APPLY DEMAND FACTORS <NEC References - 220.42 and Table 220.42>

a. First 3000VA or less of above TOTAL (at 100 percent)   = _____ VA

b. _____ x .35   = _____ VA
       (Total VA – 3001VA up to 117,000VA)

c. _____ x .25   = _____ VA
       (Remainder of TOTAL VA exceeding 120,000VA)*

*(_____ VA (TOTAL) – 120,000 VA = _____ VA)

TOTAL (Lines **a.** - **c.**)   = _____ VA
(Derated General Lighting and Receptacle Loads,   (Enter value below -
Small Appliance and Laundry Circuit Loads)   LINE and NEUTRAL)

GENERAL LIGHTING and RECEPTACLE, SMALL APPLIANCE and LAUNDRY LOADS

1. - 3. <u>LINE LOAD</u>   <u>NEUTRAL LOAD</u>

_____ VA   _____ VA

4.   APPLIANCE LOADS (Fastened-In-Place) <NEC Reference - 220.53> (Use nameplate rating of each appliance. Electric ranges, dryers, space-heating equipment or air-conditioning equipment not included.)

120V Appliances	VA		No. of appliances		VA (Total)
1. _____	_____	x	_____	=	_____
2. _____	_____	x	_____	=	_____
3. _____	_____	x	_____	=	_____
4. _____	_____	x	_____	=	_____
5. _____	_____	x	_____	=	_____
6. _____	_____	x	_____	=	_____
7. _____	_____	x	_____	=	_____
8. _____	_____	x	_____	=	_____
9. _____	_____	x	_____	=	_____
10. _____	_____	x	_____	=	_____

(Total 120V Appliances) _____

208V or 240V Appliances	VA		No. of appliances		VA (Total)
1. _____	_____	x	_____	=	_____
2. _____	_____	x	_____	=	_____
3. _____	_____	x	_____	=	_____
4. _____	_____	x	_____	=	_____
5. _____	_____	x	_____	=	_____
6. _____	_____	x	_____	=	_____
7. _____	_____	x	_____	=	_____
8. _____	_____	x	_____	=	_____
9. _____	_____	x	_____	=	_____
10. _____	_____	x	_____	=	_____

(Total 208V or 240V Appliances) _____

APPLIANCES TOTAL (120V and 208V or 240V) = _____ VA

APPLY DEMAND FACTOR (Applicable, when number of above appliances exceeds four (4) or more.)

Appliances Total VA x .75 = _____ VA

APPLIANCE LOADS

4.   <u>LINE LOAD</u>   <u>NEUTRAL LOAD (Refer to Conditions)</u>

_____ VA   _____ VA

Conditions
When appliance loads are line-to-neutral (common) [240/120V (208/120V)] or neutral (120V) connections, NEUTRAL LOAD is same as LINE LOAD.
-or-
When appliances are line-to-line (208V or 240V) loads NEUTRAL LOAD equals,

$$\underbrace{\rule{2cm}{0.4pt}}_{\text{(Line Load)}} - \underbrace{\rule{3cm}{0.4pt}}_{\text{(Line-to-Line Loads)}} \times .75^+ = \underbrace{\rule{3cm}{0.4pt}}_{\text{(NEUTRAL LOAD)}}$$

$^+$Demand Factor only applied when difference still results to 4 or more appliances.

5. CLOTHES DRYERS <NEC References - 220.54 and Table 220.54> (Use 5000W (VA) or nameplate rating, whichever is larger). Refer to, **(Worksheet E [Volume 4])** Demand Load Calculations for Household Electric Dryers. *Applies only when dryers are supplied by 240/120V.

CLOTHES DRYERS

5.   <u>LINE LOAD</u>   <u>NEUTRAL LOAD</u>
                                    <NEC Reference - 220.61(B)(1)* or (C)(1)>

_____ VA   _____ VA
                                      *(70 percent of line load)

6. ELECTRIC RANGES or OTHER COOKING APPLIANCES <NEC References - 220.55 and Table 220.55> Refer to, **(Worksheet F [Volume 4])** Demand Load Calculations for Household Electric Ranges and Other Cooking Appliances. Line-to-line electric range or other cooking appliance loads must not be included in neutral load calculations. *Applies only when electric range or other cooking appliances are supplied by 240/120V.

RANGE/COOKING APPLIANCES

6.   <u>LINE LOAD</u>   <u>NEUTRAL LOAD</u>
                                      <NEC Reference - 220.61(B)(1)* or (C)(1)>

_____ VA   _____ VA
                                      *(70 percent of line load)

7. HEATING and AIR-CONDITIONING (AC) EQUIPMENT <NEC References - 220.50, 220.51*, 220.60, Table 430. 248 and 440.6(A)> (Include VA rating of air handler [blower]). If heat pump, add compressor load and the maximum amount of electric heat that will be energized simultaneously [at the same time].) Apply, if Heat Pump is used entirely for heating and air-conditioning load. If not, use larger of Heating and Air-Conditioning loads. Larger load to include all loads that could operate at the same time, *example*: where Heat Pump is used for supplemental needs add to larger load if applicable. Use the following formulas to perform required load calculations.

Heating Load

(1) Electric Heat - _____ W (VA) x _____ x _____ = _____ VA
$\quad\quad\quad\quad\quad\quad$ (Unit Rating)$^+$ $\quad\quad\quad\quad\quad$ (Percentage)* $\quad$ (No.)*** $\quad$ (Total)

(2) Electric Heat - _____ W (VA) x _____ x _____ = _____ VA
$\quad\quad\quad\quad\quad\quad$ (Unit Rating)$^+$ $\quad\quad\quad\quad\quad$ (Percentage)* $\quad$ (No.)*** $\quad$ (Total)

(3) Electric Heat - _____ W (VA) x _____ x _____ = _____ VA
$\quad\quad\quad\quad\quad\quad$ (Unit Rating)$^+$ $\quad\quad\quad\quad\quad$ (Percentage)* $\quad$ (No.)*** $\quad$ (Total)

$^+$If calculation required, use the following formula. $\quad$ TOTAL HEAT = _____ VA
(V x A) = Unit Rating $\quad\quad\quad\quad\quad\quad\quad\quad\quad\quad\quad\quad\quad\quad$ (Combined)

AC Load

(1) AC - _____ VA^a + _____ VA^b x _____ = _____ VA
$\quad\quad\quad$ (Voltage x amps**) $\quad$ (Voltage x amps**) $\quad$ (No.)*** $\quad$ (Total)

(2) AC - _____ VA^a + _____ VA^b x _____ = _____ VA
$\quad\quad\quad$ (Voltage x amps**) $\quad$ (Voltage x amps**) $\quad$ (No.)*** $\quad$ (Total)

(3) AC - _____ VA^a + _____ VA^b x _____ = _____ VA
$\quad\quad\quad$ (Voltage x amps**) $\quad$ (Voltage x amps**) $\quad$ (No.)*** $\quad$ (Total)

$\quad\quad\quad\quad\quad\quad\quad\quad\quad\quad$ TOTAL AC = _____ VA
$\quad\quad\quad\quad\quad\quad\quad\quad\quad\quad\quad\quad\quad\quad\quad\quad$ (Combined)

Heat Pump Load

(1) Heat Pump - _____ VA^a + _____ VA^b +
$\quad\quad\quad\quad\quad\quad$ (Voltage x amps**) $\quad\quad$ (Voltage x amps**)
$\quad\quad\quad\quad\quad$ _____ W(VA) x _____ = _____ VA
$\quad\quad\quad\quad\quad$ (Maximum Electric Heat) $\quad$ (No.)*** $\quad$ (Total)

(2) Heat Pump - _____ VA^a + _____ VA^b +
$\quad\quad\quad\quad\quad\quad$ (Voltage x amps**) $\quad\quad$ (Voltage x amps**)
$\quad\quad\quad\quad\quad$ _____ W(VA) x _____ = _____ VA
$\quad\quad\quad\quad\quad$ (Maximum Electric Heat) $\quad$ (No.)*** $\quad$ (Total)

(3) Heat Pump - _____ VA^a + _____ VA^b +
$\quad\quad\quad\quad\quad\quad$ (Voltage x amps**) $\quad\quad$ (Voltage x amps**)
$\quad\quad\quad\quad\quad$ _____ W(VA) x _____ = _____ VA
$\quad\quad\quad\quad\quad$ (Maximum Electric Heat) $\quad$ (No.)*** $\quad$ (Total)

$\quad\quad\quad\quad\quad\quad\quad\quad$ TOTAL HEAT PUMP = _____ VA
$\quad\quad\quad\quad\quad\quad\quad\quad\quad\quad\quad\quad\quad\quad\quad\quad\quad\quad$ (Combined)

^{a}Compressor $\quad\quad$ ^{b}Fan Motor
$\quad$**RLA (running load amps) or FLC (full-load current) $\quad$***Number of units with identical operating characteristics.

LINE LOAD = _____ VA + _____ VA + _____VA
                (Larger Load)   (Other[s])$^{\bullet}$ (Total)

$^{\bullet}$Blowers/Air Handlers, heat pump(s) for supplement needs, etc. – if applicable. (Use [V x A] when VA not given).

HEATING and AC

7.   __LINE LOAD__   __NEUTRAL LOAD__ (120V motors only)

     _____ VA   _____ VA

8. LARGEST MOTOR <NEC References - 430.17*, 430.24, 440.7* and 440.33> (Use motor with highest current per NEC 430.17 and 440.7)

a. _____
     *([Identify] Largest Motor based on highest FLC)

b. _____ VA x .25 = _____VA
     (Largest Motor)

LARGEST MOTOR

8.   __LINE LOAD__   __NEUTRAL LOAD__ (120V motors only)

     _____ VA   _____ VA

TOTAL DEMAND LOAD (LINE and NEUTRAL) (List each computed line and neutral loads below and total lines 1. - 8.)

	LINE LOAD	NEUTRAL LOAD
1. - 3. General Lighting and Receptacles Loads, Small Appliances and Laundry Circuit Loads	_____	_____
4. Appliances	_____	_____
5. Clothes Dryer	_____	_____
6. Electric Range or Other Cooking Appliances	_____	_____
7. Heating and Air-Conditioning (AC) Equipment	_____	_____
8. Largest Motor	_____	_____
TOTAL DEMAND LOAD (VA) =	_____	_____

DWELLING'S OPERATING LINE VOLTAGE - _____ V(1φ) _____ V(3φ)
(Given operating voltage or as determined per test examination)

CALCULATE MINIMUM LINE (SERVICE) and NEUTRAL LOADS
(Divide Total Demand Load [VA] by operating line voltage [V])

LINE LOAD = _____ **VA** / (_____ **V** [x 1.732, If 3ϕ]) = _____ **A**
NEUTRAL LOAD* = _____ **VA** / (_____ **V** [x 1.732, If 3ϕ]) = _____ **A**

*Where the feeder or service neutral load exceeds 200A, NEC 220.61(B)(2) permits the load to be reduced by 70 percent. However, this reduction is not permitted if the feeder or service neutral load consist of nonlinear loads supplied from a 4-wire, wye-connected, 3-phase system (3ϕ-208/120V) *or* if supplied from any portion of a 3-wire circuit consisting of 2 ungrounded conductors (1ϕ-208/120V) and the neutral of a 4-wire, 3-phase, wye-connected system (3ϕ-208/120V). Complete the following to determine the permitted Neutral Load.

(1) _____ A – 200A = _____ x .70 = _____ A
      (Neutral Load)   (Remainder)   (Permitted Reduction)

(2) _____ A + 200A = _____ A
      (Permitted Reduction)   (Permitted Neutral Load)

SIZE SERVICE (Size of service based on calculated LINE LOAD)

SIZE SERVICE REQUIRED (minimum) _____A

SIZING FEEDER/SERVICE CONDUCTORS - (Based on the calculated LINE LOAD and NEC References.) <NEC References - 215.2(A), 230.42(A), 240.4(B) & (C), 310.10(H), 310.15(B)(2) & (3), and Table 310.15(B)(16)>

FEEDER/SERVICE CONDUCTORS _____

SIZING NEUTRAL CONDUCTOR - (Based on the calculated NEUTRAL LOAD and NEC References.) <NEC References - 215.2(A)(2), 220.61, 230.42(C), 250.24(C), 310.10(H), 310.15(B)(2), (3) & (5), 310.15(B)(7) and Table 310.15(B)(16)>

NEUTRAL CONDUCTOR(S) _____

SIZING GROUNDING ELECTRODE CONDUCTOR <NEC References - 250.24(C), 250.66 & Table 250.66 and Table 8 of Chapter 9>

GROUNDING ELECTRODE CONDUCTOR _____

**OPTIONAL LOAD CALCULATIONS
FOR MULTIFAMILY DWELLING
<NEC Reference 220.84 - Part IV>**

The Optional Load Calculation for a **Multifamily Dwelling** is only permitted for use when a multifamily dwelling will supply three or more dwelling units of a multifamily dwelling by means of a feeder or service in accordance with Table 220.84. Where Table 220.84 is applied instead of Part III of Article 220 the following conditions must be met: **(1)** No dwelling unit is supplied by more than one feeder, *****(2)** Each dwelling unit is equipped with electric cooking equipment, *and* **(3)** Each dwelling unit is equipped with either electric space heating or air conditioning, or both. Because this optional method does not include provisions for calculating the neutral load the standard method must be applied in accordance with NEC 220.61.

*Exception: When the calculated load for multifamily dwellings without electric cooking in Part III of Article 220 exceeds that calculated under Part IV for the identical load plus electric cooking (based on 8kW per unit), the lesser of the two loads shall be permitted to be used.

NEC 220.84(C) – Calculated Loads (1. - 5.)

1. **GENERAL LIGHTING and RECEPTACLE LOADS** <NEC Reference - 220.84(C)(1)> (Where multiple floor plans exist apply formula accordingly.)

$$\underline{\hspace{4cm}}\text{SF x } \underline{\hspace{2cm}} \text{ x 3VA} = \underline{\hspace{2cm}}\text{VA}$$
$$\text{(Dwelling's outside dimensions)} \quad \text{(No. of units)} \qquad \textbf{(1)}$$

2. **SMALL APPLIANCES and LAUNDRY CIRCUIT LOAD** <NEC Reference - 220.84(C)(2)> (At least two (2) small appliance and one (1) laundry branch circuit must be included.)

$$1500 \text{ VA x} \underline{\hspace{4cm}} = \underline{\hspace{2cm}} \text{ VA}$$
$$\text{(No less than 3 circuits)} \qquad \textbf{(2)}$$

3. **APPLIANCES and MOTORS LOADS** <NEC References - 220.84(C)(3) and (4)> (Use the nameplate rating of all appliances that are fastened in place, permanently connected, or located to be on a specific circuit. This includes: ranges, wall-mounted ovens, counter-mounted cooking units, clothes dryers that are not connected to the laundry branch circuit as required (per item **2**.), water heaters *and* the nameplate ampere or kVA rating of all permanently connected motors not included in this item **(3)**. **NOTE:** Heating and Air-conditioning equipment is not included.)

List all appliances and motors with VA rating. Where name plate rating is listed in amperes, use either formulas: V(volts) x A(amperes) = **VA or VA = kVA / 1000.**

Appliances and Motors	VA rating		Number of units		Total VA
Cooktops	_____	x	_____	=	_____
Dishwashers	_____	x	_____	=	_____

Disposals	_____	x	_____	= _____
Dryers	_____	x	_____	= _____
Freezers	_____	x	_____	= _____
Microwave ovens	_____	x	_____	= _____
Ranges	_____	x	_____	= _____
Refrigerators	_____	x	_____	= _____
Trash compactors	_____	x	_____	= _____
Wall-mounted ovens	_____	x	_____	= _____
Water Heaters	_____	x	_____	= _____
_____	_____	x	_____	= _____
_____	_____	x	_____	= _____
_____	_____	x	_____	= _____
_____	_____	x	_____	= _____
_____	_____	x	_____	= _____
_____	_____	x	_____	= _____
_____	_____	x	_____	= _____
_____	_____	x	_____	= _____
_____	_____	x	_____	= _____

Total Appliances VA rating = _____

(3)

4. AIR CONDITIONING or FIXED ELECTRIC SPACE-HEATING LOADS (Larger)
<NEC Reference - 220.84(C)(5)>

Air Conditioning Load = _____
Heating Load = _____

AC or Heat	VA rating	x	Number of units		Total VA
_____	_____	x	_____	=	_____

(4)

5. APPLY DEMAND FACTORS <NEC References - 220.84(A) and Table 220.84>

TOTAL CALCULATED (CONNECTED) LOAD = _____ VA

[Add loads **(1)** - **(4)**]

_____ = _____

(No. of units)   (Demand Factor [%])

TOTAL CALCULATED LOAD x _____ = _____

Demand Factor (%)   Total Demand Load

DWELLING'S OPERATING LINE VOLTAGE - _____ V
(Given operating voltage or as determined per test examination)

CALCULATE MINIMUM LINE LOAD
(Divide Total Demand Load **[VA]** by operating line voltage **[V]**)

LINE LOAD = _____ **VA** / _____ **V** = _____ **A**

SIZING FEEDER/SERVICE CONDUCTORS - (Based on the calculated LINE LOAD and NEC References.) <NEC References - 215.2(A), 220.84, 230.42(A), 240.4(B) & (C), 310.10(H), 310.15(B)(2) & (3) and Table 310.15(B)(16)>

FEEDER/SERVICE CONDUCTORS _____

SIZING NEUTRAL CONDUCTOR - There is no optional load calculation procedure for sizing the neutral conductor. To size the neutral conductor use standard load calculation for multifamily dwelling or size the neutral conductor the same as feeder/service [ungrounded] conductors. <NEC References - 215.2(A)(2), 220.61, 220.84(A)(3), 230.42(C), 250.24(C), 310.10(H), 310.15(B)(2), (3) & (5), 310.15(B)(7) and Table 310.15(B)16 > As an additional reference, refer to (**Worksheet C** - Volume 4) STANDARD LOAD CALCULATIONS FOR MULTIFAMILY DWELLING.

NEUTRAL CONDUCTOR _____

SIZING GROUNDING ELECTRODE CONDUCTOR <NEC References - 250.24(C), 250.66 & Table 250.66 and Table 8 of Chapter 9>

GROUNDING ELECTRODE CONDUCTOR _____

DEMAND LOAD CALCULATIONS
FOR HOUSEHOLD ELECTRIC DRYERS
<NEC Reference 220.54>

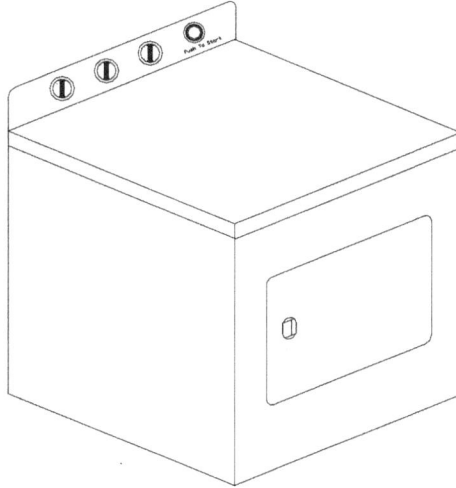

Electric Dryer

A. TO DETERMINE DEMAND LOAD WHEN SINGLE-PHASE FEEDER or SERVICE UTILIZED

1. Refer to NEC 220.54. Dryer rating must be 5000 watts (volt-amperes) *or* the nameplate rating, whichever is larger.

2. Refer to Table 220.54 to determine the dryer(s) demand factor.

   a. Where single dryer or dryers of the same rating exist apply the following formula:

   Multiply the kW (kVA) rating of the dryer(s) by the number (No.) of dryers by the demand factor to determine the demand load of the dryer(s).

   _____ kW (kVA) x _____ x _____ = _____ kW (kVA)
   (Rating)   (No. of dryers) (Demand Factor)* (Demand Load)

   b. Where dryers of different ratings exist, total the ratings of the dryers and multiply by the demand factor based upon the number of dryers to determine the demand load of the dryers.

   _____ kW (kVA) x _____ = _____ kW (kVA)
   (Total rating)   (Demand Factor - per no. of dryers)* (Demand Load)

   *Where the No. of dryers exceeds eleven (11), refer to step 3. for the actual equations of the given formulas in Table 220.54.

3. Determine the demand factor (percent) per number of dryers per Table 220.54.

    a. If the number of dryers are between **12** and **23**, use the following formula to derive the demand factor:

$$47\% \,(.47) - [1\% \,(.01) \times (\underline{\hspace{3cm}} - 11)] = \underline{\hspace{3cm}}$$
$$\text{(No. of dryers)}\text{(Demand Factor)}$$

    b. If the number of dryers are between **24** and **42**, use the following formula to derive the demand factor:

$$35\% \,(.35) - [.5\% \,(.005) \times (\underline{\hspace{3cm}} - 23)] = \underline{\hspace{3cm}}$$
$$\text{(No. of dryers)}\text{(Demand Factor)}$$

    c. If the number of dryers are over **43**, use the following demand factor:

$$25\% \,(.25)$$

B. TO DETERMINE TOTAL DEMAND FOR THREE-PHASE 4W FEEDER OR SERVICE (for two *or* more single phase dryers supplied by a 3-phase, 4-wire feeder or service [208/120V / 3-phase / 4W-Wye (Y) connected] - Based on NEC 220.54)

1. Divide the number of dryers by the number of phases.

$$\frac{\text{No. of dryers}}{\text{Phases}} = \frac{}{3} = \frac{}{\text{(Maximum No. of dryers)}} *$$

* - maximum number of dryers between two phases (must be rounded up to the nearest whole number)

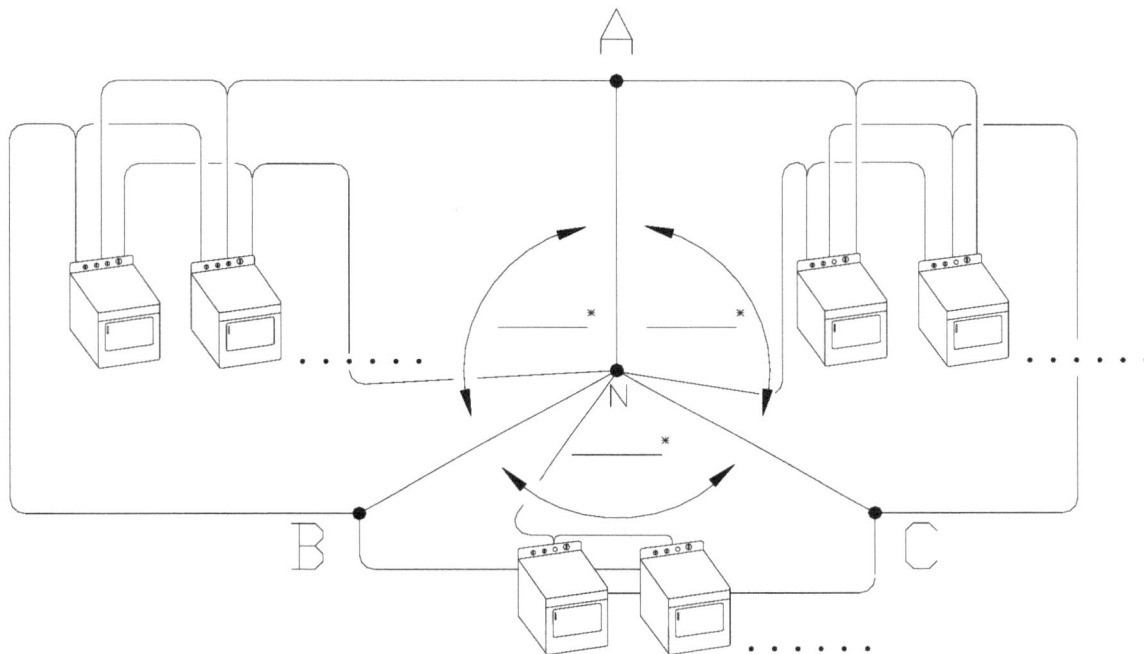

2. Multiply the maximum number of dryers times 2 (between 2 phases).

$$\frac{}{\text{(Maximum No. of dryers)}} \times 2 = \frac{}{\text{(Maximum No. of dryers between 2 phases)}}$$

3. Determine the demand factor (percent) per number of dryers per Table 220.54.

   a. If the number of dryers are between **12** and **23**, use the following formula to derive the demand factor:

   $$47\% (.47) - [1\% (.01) \times (\underline{}_{\text{(No. of dryers)}} - 11)] = \underline{}_{\text{(Demand Factor)}}$$

   b. If the number of dryers are between **24** and **42**, use the following formula to derive the demand factor:

   $$35\% (.35) - [.5\% (.005) \times (\underline{}_{\text{(No. of dryers)}} - 23)] = \underline{}_{\text{(Demand Factor)}}$$

c. If the number of dryers exceeds **43**, use the following demand factor:

25% (.25)

4. Determine the demand load per 2 phases based on the maximum number (No.) of dryers (between two phases).

_____ kW (kVA) x _____ x _____ = _____ kW (kVA)
(Rating) **   (No. of dryers)   (Demand Factor)   (Demand Load per 2 phases)

    **Rating must be 5000 watts (volt-amperes) *or* the nameplate rating, whichever is larger.

5. Determine single phase **(1ϕ)** and three phase **(3ϕ)** Demand Loads.

$$\frac{\text{Demand Load per 2 phases}}{2} = \text{____VA (kW) [Demand Load per phase (1ϕ)]}$$

VA (kW) [Demand Load per phase **(1ϕ)**] x 3 = _____3 phase **(3ϕ)** Demand Load

6. Use either of the following formulas to determine the load in amperes (A).

$$\frac{\text{Demand load per phase (1ϕ)}}{120V} \quad or \quad \frac{\text{3 phase (3ϕ) Demand Load}}{208V \times 1.732}$$

DEMAND LOAD CALCULATIONS
FOR HOUSEHOLD ELECTRIC RANGES
AND OTHER COOKING APPLIANCES
<NEC Reference 220.55>

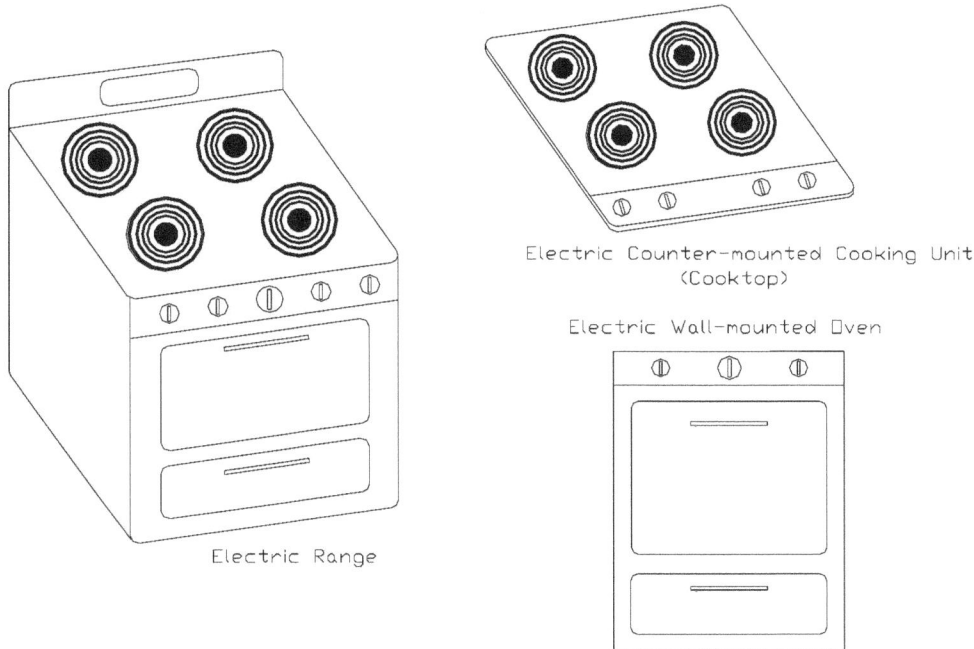

Electric Counter-mounted Cooking Unit
(Cooktop)

Electric Wall-mounted Oven

Electric Range

TABLE 220.55 - Household Electric Ranges, Wall-Mounted Ovens, Counter-Mounted Cooking Units, and Other Household Cooking Appliances over 1¾(1.75)kW Rating

Column A - Covers appliances that are rated between 1.76kW (over 1¾kW) and 3.49kW (less than 3½kW).

Column B - Covers appliances that are rated between 3½ (3.5)kW and 8¾ (8.75)kW.

The numbers listed under Columns A and B are percentage values (demand factors) based on the number of appliances being used. These percentages values are used to reduce the total kilowatt rating of all included appliances per column.

Column C - Covers those appliances rated between 8.76kW (over 8¾kW) and 12kW. In general Column C covers all appliances rated over 1¾ (1.75)kW up to 12kW.

The numbers listed under Column C are the maximum (actual) demand based on the number of appliances being used and *requires no calculations*. Because Column C covers all appliances rated up to 12kW, those appliances with ratings that are applicable to Columns A and B falls within the boundaries of Column C also. Therefore, all calculated demand loads derived from Columns A *or* B should be compared to the maximum demand in Column C to determine the lowest (minimum) demand load to be used. Such use is permitted at the conclusion (parenthesis) of the heading in Table 220.55 which implicitly permits the use of the lowest demand load.

Column A - Appliances rated between 1.76kW and 3.49 kW

TO DETERMINE DEMAND LOAD

1. List the number of appliances: _____

2. Total the kW rating of appliances: _____ kW (Total kW)

3. Multiply the Total kW by the demand factor (percent) listed in **Column A** per number of appliances to determine the demand load of the appliance(s).

$$\underset{\text{(Total kW)}}{\underline{\hspace{2cm}}} \text{kW} \ \times \ \underset{\text{(Demand Factor)}}{\underline{\hspace{3cm}}} = \underset{\text{(Demand Load)}}{\underline{\hspace{2.5cm}}} \text{kW}$$

4. Compare the calculated demand load per **Column A** with the Maximum Demand of **Column C** per number of appliances. Use the lowest demand load.

$$\underset{\textbf{(Column A} \text{ - demand load)}}{\underline{\hspace{3cm}}} \text{kW} \qquad \underset{\textbf{(Column C} \text{ - Maximum Demand)}}{\underline{\hspace{4cm}}} \text{kW}$$

Column B - Appliances rated between 3.5kW and 8¾(8.75)kW

TO DETERMINE DEMAND LOAD

1. List the number of appliances: _____

2. Total the kW rating of appliances: _____ kW (Total kW)

3. Multiply the Total kW by the demand factor (percent) listed in **Column B** per number of appliances to determine the demand load of the appliance(s).

$$\underset{\text{(Total kW)}}{\underline{\hspace{2cm}}} \text{kW} \ \times \ \underset{\text{(Demand Factor)}}{\underline{\hspace{2.5cm}}} = \underset{\text{(Demand Load)}}{\underline{\hspace{2cm}}} \text{kW}$$

4. Compare the calculated demand load per **Column B** with the Maximum Demand of **Column C** per number of appliances. Use the lowest demand load.

$$\underset{\textbf{(Column B} \text{ - demand load)}}{\underline{\hspace{3cm}}} \text{kW} \qquad \underset{\textbf{(Column C} \text{ - Maximum Demand)}}{\underline{\hspace{4cm}}} \text{kW}$$

Column C - Appliances rated from 1.76kW to 12kW

TO DETERMINE DEMAND LOAD

1. Use the Maximum Demand listed in **Column C** per number of appliances.

2. Where the number of appliances exceeds 25, the Maximum Demand must be determined based on the following formulas:

   a. If the number of appliances are between **26** and **40**, use the given formula to derive the Maximum Demand:

$$15\text{kW} + (1\text{kW x} \underline{\hspace{2cm}}) = \underline{\hspace{2cm}}$$
$$\qquad\qquad\qquad (\text{No. of ranges})\qquad (\text{Maximum Demand})$$

   b. If the number of appliances are over **40**, use the given formula to derive the Maximum Demand:

$$25\text{kW} + ([¾] .75\text{kW x} \underline{\hspace{2cm}}) = \underline{\hspace{2cm}}$$
$$\qquad\qquad\qquad\quad (\text{No. of ranges})\qquad (\text{Maximum Demand})$$

TABLE 220.55, Notes 1 - 4

Note 1 - (Over 12kW thru 27kW ranges - Same Rating)

TO DETERMINE DEMAND LOAD*

1. $\underline{\hspace{3cm}}$ $- 12\text{kW} =$ $\underline{\hspace{3cm}}$
   (Range Nameplate Rating [kW])   (Additional kilowatts [kW])

2. $\underline{\hspace{3cm}}$ x .05 $=$ $\underline{\hspace{1.5cm}}$ + 1 = $\underline{\hspace{3cm}}$
   (Additional kilowatts [kW])   (Percent)   [Percent Multiplier (PM)]**

3. $\underline{\hspace{4cm}}$ x $\underline{\hspace{1cm}}$ $=$ $\underline{\hspace{2.5cm}}$
   (Max. Demand [**Col. C**] per No. of Appls.)   (PM)   (Demand Load)

Note 2 - (Over 8¾ [8.75]kW thru 27kW ranges - Unequal Rating)

TO DETERMINE DEMAND LOAD*

1. (Total kW Rating of all Ranges)*** $=$ $\underline{\hspace{2.5cm}}$ *
   (Total No. of Ranges)   (Average Rating)

2. $\underline{\hspace{2.5cm}}$ $- 12\text{kW} =$ $\underline{\hspace{3cm}}$
   (Average Rating)   (Additional kilowatts [kW])

3. $\underline{\hspace{3cm}}$ x .05 $=$ $\underline{\hspace{1.5cm}}$ + 1 = $\underline{\hspace{3cm}}$
   (Additional kilowatts [kW])   (Percent)   [Percent Multiplier (PM)]**

4. _____ x _____ = _____
(Max. Demand [**Col. C**] per No. of Appls.)   (PM)   (Demand Load)

   * - Major Fraction (**Notes 1** and **2**). When the kilowatt (kW) rating of an appliance involves a fractional (decimal) value and that value is 0.5 and greater, round the appliance rating up to the nearest whole number - otherwise round down <u>before proceeding</u>.
   ** - Percent multiplier (short-cut method) allows results to be derived in one-step.
   *** - Use 12kW where rating of range is either equal to *or* greater than 8.75kW and less than 12kW.

Note 3 - (Over 1¾[1.75]kW thru 8¾[8.75]kW appliances)

In lieu (instead) of using the maximum demand values given in Column C of Table 220.55, **Note 3** permits adding the nameplate ratings of all household cooking equipment rated more than 1¾kW but not more 8¾kW and multiplying the sum by the demand factors specified in Columns A *or* B based on the given number of appliances. Afterwards, the results can then be compared with the applicable Maximum Demand of Column C.

TO DETERMINE DEMAND LOAD

1. List the number of appliances rated between **1.76kW** and **3.49 kW**.
   _____ _____ _____ _____ _____ _____ _____

2. Total the kW rating of appliances: _____ kW (Total kW)

3. Multiply the Total kW by the demand factor (percent) listed in **Column A** per number of appliances to determine the demand load of the appliance(s).

   _____ kW x _____ = _____ kW
   (Total kW)   (Demand Factor)   (Demand Load A)

4. List the number of appliances rated between **3.5kW** and **8.75 kW**.
   _____ _____ _____ _____ _____ _____ _____

5. Total the kW rating of appliances: _____ kW (Total kW)

6. Multiply the Total kW by the demand factor (percent) listed in **Column B** per number of appliances to determine the demand load of the appliance(s).

   _____ kW x _____ = _____ kW
   (Total kW)   (Demand Factor)   (Demand Load B)

7. Total the demand loads.

   _____ + _____ = _____
   (Demand Load A)   (Demand Total B)   (Total Demand Load)

8. Compare the Total Demand Load with the Maximum Demand of **Column C** based on the total number of appliances. Use the lowest demand load.

_____ kW   _____ kW
(Total Demand Load)   (**Column C** - Maximum Demand)

Table 220.55, Note 4

TO DETERMINE BRANCH CIRCUIT LOAD

1. For one (1) range apply the appropriate provisions per Columns A, B or C to derive load.

   Branch-Circuit Load = Derived Load / 240V (or voltage provided)

2. For one (1) wall-mounted oven _or_ one (1) counter-mounted cooking unit, branch circuit load _based on_ the **nameplate rating** of appliance.

   Branch-Circuit Load = Nameplate Rating / 240V (or voltage provided)

3. For one (1) counter-mounted cooking unit and _not more than two (2)_ wall-mounted oven units where all appliances are supplied from a single branch circuit and located in the same room, total the nameplate rating of each appliance and <u>treat the total as equivalent to one range</u>.

   a. If total (nameplate ratings) is less than 12kW, apply the applicable Column's demand factor _or_ maximum demand. Compare and use the lowest value.

      Branch-Circuit Load = Lowest value / 240V (or voltage provided)

   b. If total (nameplate ratings) is greater than 12kW, apply **Note 1.**

      Branch-Circuit Load = Demand Load / 240V (or voltage provided)

<u>Single phase ranges supplied by 3-phase 4W Feeder or Service</u>

TO DETERMINE TOTAL DEMAND FOR 3-PHASE 4W FEEDER OR SERVICE (for two _or_ more single phase ranges supplied by a 3-phase, 4-wire feeder or service [208/120V / 3-phase / 4W-Wye (Y) connected] - Based on NEC 220.55)

1. Divide the number of ranges by the number of phases.

   $$\frac{\text{No. of ranges}}{\text{Phases}} = \frac{}{3} = \underline{\hspace{1cm}} *$$

   * - maximum number of ranges between two phases (must be rounded up to the nearest whole number)

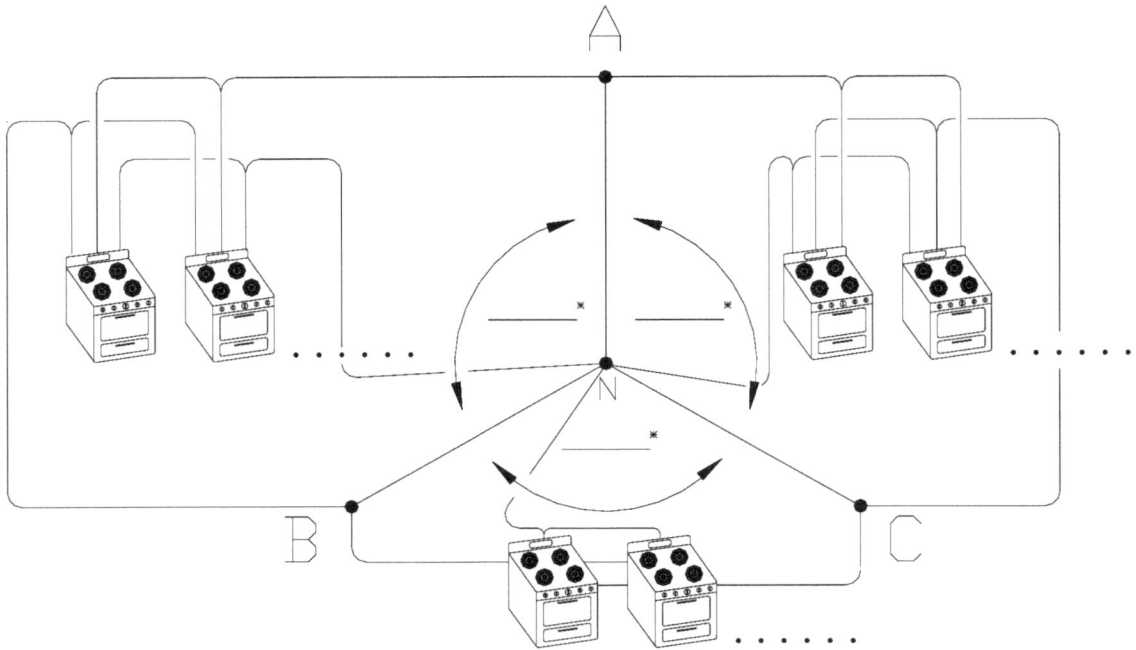

2. Multiply the maximum number of ranges times 2 (between 2 phases).

$$\underline{\hspace{4cm}} \text{ x } 2 = \underline{\hspace{6cm}}$$
(No. of ranges per phase)   (Maximum No. of ranges between 2 phases)

3. Determine the demand load per 2 phases per Columns A, B, C *or* Table 220.55, Note 1 based on the maximum number (No.) of ranges (between two phases).

    Column A (Ranges less than 3½kW) *or* **B** (Ranges between 3½kW - 8¾kW)

$$\underline{\hspace{3cm}} \text{ kW x } \underline{\hspace{3cm}} \text{ x } \underline{\hspace{3cm}} = \underline{\hspace{4cm}}$$
(range rating)   (No. of ranges)   (Demand Factor)   (Demand Load per 2 phases)
                                                  [per **Col. A** *or* **B**]

Per Column **A** *or* **B**, compare the results of the demand load of the ranges with the maximum demand in **Column C**.

Column C

    Maximum demand (Demand Load per 2 phases) = $\underline{\hspace{5cm}}$
                                                  (Based on No. of ranges)

Note 1 (Ranges over 12kW through 27kW - same rating)

a. $\underline{\hspace{6cm}}$ − 12kW = $\underline{\hspace{5cm}}$
   (Range Nameplate Rating [kW])   (Additional kilowatts [kW])

b. _____ x .05 = _____ + 1 = _____
    (Additional kilowatts [kW])   (Percent)   (PM)

c. _____ x _____ = _____
    (Max. Demand [Col C] per No. of ranges.)   (PM)   (Demand Load per 2 phases)

4. Determine single phase **(1ϕ)** and three phase **(3ϕ)** Demand Loads.

$$\frac{\text{Demand Load per 2 phases}}{2} = \underline{\quad} \text{ VA (kW) [Demand Load per phase } \textbf{(1ϕ)]}$$

VA (kW) [load per phase **(1ϕ)**] x 3 = _____ 3 phase **(3ϕ)** Demand Load

5. Use either of the following formulas to determine the load in amperes (A).

$$\frac{\text{Demand load per phase } \textbf{(1ϕ)}}{120V} \quad or \quad \frac{\text{3 phase } \textbf{(3ϕ)} \text{ Demand Load}}{208V \times 1.732}$$

STANDARD LOAD CALCULATIONS for NONDWELLING BUILDINGS (COMMERCIAL and INDUSTRIAL)

1. GENERAL LIGHTING <NEC References - 220.12 and Table 220.12> or **ACTUAL LIGHTING LOADS** (Apply the larger lighting load)

FN = Footnote $\sqrt{3} = 1.732$

General Lighting Load (Square feet [SF] per occupancy – VA/SF [Table 220.12])

_____ - _____ SF x ___ VA/SF = _____ VA
(Type Occupancy)

_____ - _____ SF x ___ VA/SF = _____ VA
(Type Occupancy)

_____ - _____ SF x ___ VA/SF = _____ VA
(Type Occupancy)

_____ - _____ SF x ___ VA/SF = _____ VA
(Type Occupancy)

_____ - _____ SF x ___ VA/SF = _____ VA
(Type Occupancy)

A. General Lighting Load [FN1] = _____ VA

Actual Lighting Load (List all fixtures that will contribute to the occupancy's entire interior lighting load)

Type Fixture	VA rating [FN2]	No. of Fixtures	TOTAL VA
_____	_____ x	_____ =	_____
_____	_____ x	_____ =	_____
_____	_____ x	_____ =	_____
_____	_____ x	_____ =	_____
_____	_____ x	_____ =	_____
_____	_____ x	_____ =	_____
_____	_____ x	_____ =	_____

B. Actual Lighting Load [FN1] = _____ VA
(or given Load)

[FN1] - Apply DEMAND FACTORS if HOSPITAL or WAREHOUSE based on Larger Lighting Load (**A.** or **B.**), if applicable.

[FN2] - If INCANDESCENT FIXTURE, use wattage of lamp (bulb). For lighting units that have BALLAST, TRANSFORMERS, AUTOTRANSFORMERS, or LED DRIVERS, the calculated load shall be based on the total ampere ratings of such units [NEC 220.18(B)].

APPLY DEMAND FACTORS <NEC References - 220.42 and Table 220.42>

FOR HOSPITALS ONLY*

(If Total VA is less than or equal to 50,000VA, step **b.** is not required)

a. First 50,000VA or less of lighting load (at 40%)

_____ x .40 _____ = _____ VA

b. _____ x .20 (at 20%) = _____ VA
(Remainder of TOTAL VA exceeding 50,000VA)**

**(_____ VA (TOTAL) – 50,000VA = _____ VA)

TOTAL (Lines **a.** and **b.**) = _____ VA
(Derated Lighting Load)

* - Areas where lighting will be used continuously are to be listed in **2E** - Other Lighting Loads.

FOR WAREHOUSES (STORAGE) ONLY

(If Total VA is less than or equal to 12,500VA, step **b.** is not required)

a. First 12,500VA or less of lighting load (at 100%) = _____ VA

b. _____ x .50 (at 50%) = _____ VA
(Remainder of TOTAL VA exceeding 12,500VA)***

***(_____ VA (TOTAL) – 12,500VA = _____ VA)

TOTAL (Lines **a.** and **b.**) = _____ VA
(Derated Lighting Loads)

_____ VA x 1.25 [FN3 and FN4] = _____ VA
Enter [Larger Lighting Load (**A.** or **B.**)]or   (LINE LOAD)
[Derated Lighting Load where applicable]
This entry, used for NEUTRAL LOAD

[FN3] - Omit 125 percent increase, if DEMAND FACTORS were applied.

[FN4] - <NEC References - 215.2(A)(1) [FEEDER] or 230.42(A)[SERVICE]> (Line loads are increased by 125 percent because they are expected to operate for 3 hours or more [continuous]. **Neutral loads** are computed at 100 percent. Although the neutral conductor(s) will serve continuous loads, under most conditions it will never experience the same demands as the line conductors and is subject to derating per NEC 220.61. For clarity, Neutral Loads as it pertains to this worksheet are recognized as those loads that are associated with, **(1)** a *neutral conductor* as described in Article 100 *or* **(2)** a *grounded conductor* which carries the same amount of current as an ungrounded conductor *or* **(3)** a *common conductor* as referenced in NEC 310.15(B)(5). See Article 310 (NEC 310.15(B)(5) [Volume 2].

1. <u>LINE LOAD</u>   [FN4] <u>NEUTRAL LOAD</u>

[FN5] Permitted   Prohibited
Reduction   Reduction

_____ VA   _____ VA   _____ VA

[FN5] - Reduction of the feeder or service neutral load is **permitted** to have an additional 70 percent (.70) applied when supplying household electric ranges, wall-mounted ovens, counter-mounted cooking units and dryers. This 70 percent (.70) reduction is also permitted for unbalanced loads in excess of 200 amperes where the feeder or service supply *linear loads** from a 3-wire dc *or* single-phase ac system, *or* a 4-wire, 3-phase; 3-wire, 2-phase-system, *or* a 5-wire, 2-phase system.

Reduction of the feeder or service neutral load is **prohibited** for that portion of the load that consist of: **(1)** a 3-wire circuit consisting of 2 ungrounded conductors (208/120V- 1φ) and the neutral of a 4-wire, 3-phase, wye-connected system (208/120V-3φ) *and* **(2)** that portion of the load consisting of *nonlinear loads*** supplied from a 4-wire, wye connected, 3-phase system (208/120V- 3φ and 480/277V- 3φ).

*Examples of *linear loads*: Heating equipment, electric motors, resistive lighting (incandescent), etc.
**Examples of *nonlinear loads*: computer equipment, converters, data-processing equipment, drives (adjust-able/frequency/speed/variable), electronic ballast, electric discharge lighting (fluorescent, high and low-pressure sodium, mercury-vapor, metal-halide, etc.), inverters, medical and laboratory test equipment, programmable logic controllers (PLC), UPS systems, welders, etc.

2. OTHER LIGHTING LOADS

A. Sign/Outline (S/O) Lighting <NEC References - 220.14(F) and 600.5(A)>

S/O 1 - _____ V x _____ A (or given VA(W) Load) = _____ VA
              (Voltage)   (Amperes)   (Minimum of 1200VA required)

S/O 2 - _____ V x _____ A (or given VA(W) Load) = _____ VA
              (Voltage)   (Amperes)   (Minimum of 1200VA required)

S/O 3 - _____ V x _____ A (or given VA(W) Load) = _____ VA
              (Voltage)   (Amperes)   (Minimum of 1200VA required)

Total (Sign/Outline Lighting) = _____ VA
                                    (2A)

B. Outside Lighting <NEC Reference - 220.18(B)>

_____ _____V x ____A or _____W(VA) x _____ = _____VA
(Type - 1) (Volts) (Amps) (Lamp Watts) (No.) [FN6] (Load)•

_____ _____V x ____A or _____W(VA) x _____ = _____VA
(Type - 2) (Volts) (Amps) (Lamp Watts) (No.) [FN6] (Load)•

_____ _____V x ____A or _____W(VA) x _____ = _____VA
(Type - 3) (Volts) (Amps) (Lamp Watts) (No.) [FN6] (Load)•

[FN6] - Number of Fixtures with identical operating characteristics. •Calculated or provided by other.

Total (Outside Lighting) = _____ VA
                               (2B)

C. Show-Window Lighting <NEC Reference - 220.43(A)> (Voltage rating _____ V)

_____ x 200VA (or given VA Load) = _____ VA
(Linear Feet)   **(2C)**

D. Track Lighting <NEC Reference - 220.43(B)> (Voltage rating _____ V)

(_____ ÷ 2') x 150VA (or given VA Load) = _____ VA
(Linear Feet)   **(2D)**

E. Miscellaneous (Write-ins. List individual voltage rating of each lighting [fixture] load.)

_____
_____
_____

Total (Miscellaneous) = _____ VA
                        (2E)

F. Other Lighting (LINE and NEUTRAL) Loads Total [Add lines **(2A)** - **(2E)** where applicable]

TOTAL = _____ VA x 1.25 [FN4] + _____ = _____ VA
        (Continuous)   (Noncontinuous)   (LINE LOAD)

2. <u>LINE LOAD</u>   [FN4] <u>NEUTRAL LOAD</u>
                                        (120V or 277V)
                                [FN5] Permitted   Prohibited
                                      Reduction   Reduction

_____ VA   _____ VA   _____ VA

3. RECEPTACLE LOADS (120 volts only)

A. Non-continuous duty <NEC References 220.14(H) and (I), 220.44 and Table 220.44>

(1) Fixed Multioutlet Assemblies [FN7] <220.14(H)>

Non-simultaneous use
(_____ ÷ 5) x 180VA = _____ VA
(Linear Feet)   **(3A)**

Simultaneous use
_____ x 180VA = _____ VA
(Linear Feet)   **(3B)**

[FN7] - Usually not considered a continuous load. However if so, include that portion with **3E** and proceed.

(2) General Purpose Receptacles* and Fixed Multioutlet Assemblies < 220.14(I)>

$$\underline{\hspace{3cm}} \times 180VA^{\centerdot} = \underline{\hspace{1.5cm}} VA + \underline{\hspace{1.5cm}} VA = \underline{\hspace{1.5cm}} VA^{[FN8]}$$
(No. of receptacles)*   **(3A) + (3B)**   (TOTAL)

*Only apply if other than bank or office building. For bank and office buildings apply item **B**.

[FN8] - APPLY DEMAND FACTORS [If applicable]

a. First 10,000VA (10kVA) of above TOTAL   = _____VA
(At 100 percent)

b. _____ x .50   = _____VA
(Remainder of TOTAL exceeding 10,000VA)**

**(_____ VA (TOTAL) – 10,000 VA = _____ VA)

_____ VA
(3C) [Total - **(2)a.** and **b.**]

B. Non-continuous duty (If Bank or Office Building [FN9]) <NEC Reference - 220.14(K)>

[FN9] General Purpose Receptacles [Use Larger of **(1)** or **(2)** - Because item **(1)** is derived from NEC 220.14(I), the provisions of NEC and Table 220.44 are permitted.]

(1) _____ x 180VA^{\centerdot} = _____ VA [FN10]
(No. of receptacles)

[FN10] - APPLY DEMAND FACTORS [if applicable]. If the results of **(1)** is larger than **(2)**, omit **B.** and include the number of receptacles with step **3.A.(2)** above if applicable. Apply demand factors if the number of receptacles and step **3.A.(1)** exceeds 10,000VA.

a. First 10,000VA (10kVA) of above TOTAL   = _____VA
(At 100 percent)

b. _____ x .50   = _____VA
(Remainder of TOTAL exceeding 10,000VA)**

**(_____ VA (TOTAL) – 10,000 VA = _____ VA)

(2) _____ x 1 VA/SF = _____ VA
Building's Square Footage

_____ VA
(3D) [Larger of Total - **(1)a.** and **b.** or **(2)**]

C. Continuous duty <NEC References - 220.14(I), 215.2(A)(1) [FEEDER] or 230.42(A) [SERVICE]

_____ x 180VA^{\centerdot} (or 90VA^{\centerdot\centerdot}) = _____ VA x 1.25 = _____ VA
(No. of receptacles)   **(3E)**   **(3F)**

*For each single/multiple receptacle on one strap. Single piece comprised of 4 or more receptacles, 90VA.

D. Total Receptacle (LINE and NEUTRAL) Load
[LINE LOAD - Add lines (3C), (3D) and (3F)] [NEUTRAL LOAD - Add lines (3C), (3D) and (3E)]

3.   <u>LINE LOAD</u>
           [FN4] <u>NEUTRAL LOAD</u>
           [FN5] Permitted   Prohibited
                   Reduction   Reduction

_____ VA
           _____ VA   _____ VA

4. KITCHEN EQUIPMENT <NEC Reference - 220.56 > [4]*(Where kitchen equipment consist of refrigeration equipment driven by a hermetic type [sealed] motor-compressor, use **(a)** the branch-circuit selection current [bcsc] {NEC 440.6(A), *Exception No. 1*} marked on the equipment. If equipment not marked with bcsc, use **(b)** rated load current marked on the nameplate of the equipment or when no rated-load current is shown on the equipment nameplate, use **(c)** the rated-load current on the compressor nameplate [NEC 440.6(A)]. When a conventional motor is used to drive a refrigeration compressor use the full-load current [FLC] per motor's horsepower rating given in the applicable Tables of Article 430.)

List each piece of kitchen equipment (KE) per voltage rating and phases. Use the given formula to perform load calculation. If VA value given computation not required.

Type KE   Calculation (Refer to **Formula 4**)

_____   _____
_____   _____
_____   _____
_____   _____
_____   _____
_____   _____

                 TOTAL = _____ VA
                          (Total)

Formula 4

_____V x _____A x _____ = _____VA = _____VA
(Volts)   (Amps)[4]*   (If 3ϕ, √3)   (Computed)   (Total)

     a. APPLY DEMAND FACTORS <NEC Reference - Table 220.56> (Based on number of units of equipment. Does not apply to space heating, ventilating, or air-conditioning equipment.)

           _____ VA x _____ = _____ VA
             (Total VA)   (Demand Factor per Total Units)   (Demand Load) **(a)**

b. DETERMINE TWO LARGEST KITCHEN LOADS

_____ VA + _____ VA = _____ VA
(Largest - 1)   (Largest - 2)   (Total) **(b)**

KITCHEN LOAD (Compare computed values in lines **(a)** and **(b)**. Use larger of two for LINE LOAD).

NEUTRAL LOAD equals,

_____ VA x _____ = _____ VA
(Total VA of 120V, 208/120V,   (Demand Factor per Total Units)   (NEUTRAL LOAD)
240/120V Units)

4.   <u>LINE LOAD</u>   [FN4] <u>NEUTRAL LOAD</u>
                                                              [FN5] Permitted   Prohibited
                                                                     Reduction   Reduction

_____ VA   _____ VA _____ VA

5. SPECIFIC LOADS (Appliances <NEC Reference 422> Office Equipment, Computer Equipment <NEC Reference 645>, etc.)

List each load per voltage rating and phases. Use the given formula to perform load calculation. If VA value given computation not required.

Type Load   Calculation (Refer to **Formula 5**)

_____   _____
_____   _____
_____   _____
_____   _____

TOTAL = _____ VA
                        (Total)

Formula 5

_____V x _____A x _____ x 1.25$^{•••}$ x _____ = _____ VA = _____ VA
(Volts)   (Amps)   (If 3φ, √3)   (No.)   (Computed)   (Total)

•••If continuous, Total only applied to LINE LOAD. NEUTRAL LINE calculated at 100 percent.

5.   <u>LINE LOAD</u>   [FN4] <u>NEUTRAL LOAD</u>
                                                              [FN5] Permitted   Prohibited
                                                                     Reduction   Reduction

_____ VA   _____ VA _____ VA

6. MOTOR LOADS <NEC Reference - 220.50 >

List each motor load. Use the given formulas to perform load calculation.

A. Continuous Duty <NEC References - Tables 430.247 - 430.250 and 440.6(A)> This section applies to those motors that could operate at a substantially constant load for an indefinitely long time. This section also applies to hermetic refrigerant motor-compressors used with refrigerant equipment. Where hermetic refrigerant motor-compressors are used with air-conditioning equipment see Item. **9 - HEATING and AIR-CONDITIONING (AC) EQUIPMENT**.

Motor Load   Calculation (Refer to **Formula 6A**)

_____   _____
_____   _____
_____   _____
_____   _____
_____   _____

TOTAL = _____ VA
(6A) (Combined)

Formula 6A

____V x ____A x _____ = _____ VA = _____ VA
(Volts) (Amps)[6]* (If 3φ, √3) (Computed/Given)   (Total)

[6]*For conventional motors use the full-load currents in Tables 430.247 - 430.250. For hermetic refrigerant (sealed) motor-compressor, use **(a)** the branch-circuit selection current (bcsc) [NEC 440.6(A), *Exception No. 1*] marked on the equipment. If equipment not marked with bcsc, use **(b)** rated load current marked on the nameplate of the equipment or when no rated-load current is shown on the equipment nameplate, use **(c)** the rated-load current on the compressor nameplate [NEC 440.6(A)].

B. Other Than Continuous Duty <NEC References - 430.22(E) and Table 430.22(E)> [NEC 430.22(E) requires the motor's nameplate current[6]** to be used for other than continuous duty. When the nameplate current is not available use the applicable full-load current listed in the Tables of Article 430. Depending upon the motor's duty cycle and time[duty] rating, the nameplate current can be either increased or decreased per Table 430.22(E)[6]***.]

Motor Load   Calculation (Refer to **Formula 6B**)

_____   _____
_____   _____
_____   _____
_____   _____
_____   _____

TOTAL = _____ VA
(6B) (Combined)

— 46 —

Formula 6B

$$\underline{\hspace{1cm}}V \times \underline{\hspace{1.5cm}}A \times \underline{\hspace{1cm}} = \underline{\hspace{2cm}}VA \times \underline{\hspace{1cm}} = \underline{\hspace{1cm}}VA$$

(Volts)   (Amps)[6]**   (If 3φ, √3)   (Computed/Given)   (%)[6]***   (Total)

C. Cranes and Hoists (Article 610) Use motor's nameplate full-load ampere [current] rating. When the nameplate current is not available use the applicable full-load current listed in the Tables of Article 430.

(1) Single Motor <NEC Reference - 610.14(E)(1)>

List each single motor load. Use the given formula to perform load calculation. If VA value given computation not required.

Motor Load   Calculation (Refer to **Formula 6C1**)

_____   _____
_____   _____
_____   _____
_____   _____

TOTAL = _____ VA

(6C) (Combined)

Formula 6C1

$$\underline{\hspace{1cm}}V \times \underline{\hspace{1.5cm}}A \times \underline{\hspace{1cm}} = \underline{\hspace{2cm}}VA = \underline{\hspace{1cm}}VA$$

(Volts)   (Amps)[6]**   (If 3φ, √3)   (Computed/Given)   (Total)

(2) Multiple Motors on Single Crane or Hoist <NEC Reference - 610.14(E)(2)>

List each crane/hoist motors. Use the given formula to perform load calculation.

Motor Load   Calculation (Refer to **Formula 6C2**)

_____   _____
_____   _____
_____   _____

TOTAL = _____ VA

(6D) (Combined)

Formula 6C2

(Largest Motor)

$$\underline{\hspace{1cm}}V \times \underline{\hspace{1.5cm}}A \times \underline{\hspace{1cm}} = \underline{\hspace{2cm}}VA$$

(Volts)   (Amps)[6]**   (If 3φ, √3)   (Computed/Given)

(Other Motors)

_____V x _____A x _____ = _____ VA
(Volts)   (Amps)[6]**   (If 3φ, √3)   (Computed/Given)

$$\text{TOTAL} = \underline{\hspace{3cm}} \text{ VA}$$
(Other Motors)

_____ VA + (_____ VA x .50 [50%]) = _____ VA
(Largest Motor)   (Other Motors)   (Combined Total)

(3) Multiple Cranes or Hoist on Common Conductor <NEC Reference 610.14(E)(3)>

List each crane/hoist motors. Use the given formula to perform load calculation. If VA value given computation not required.

Cranes or Hoist   Calculation (Refer to **Formula 6C1**)

_____   _____
_____   _____
_____   _____

$$\text{TOTAL} = \underline{\hspace{3cm}} \text{VA}$$
(Combined)

Crane/Hoist's Demand Factor <NEC Reference - Table 610.14(E)> (Limited to seven [7] cranes or hoist per common conductor)

_____
(number [No.] of cranes or hoists)

_____ VA x _____ = _____ VA
(Combined)   (Demand Factor per No. of cranes or hoists)   **(6E)** (Demand Load)

D. Elevators, Dumbwaiters, Escalators, Moving Walks, Platform and Stairway Lifts – (Article 620) These particular loads require the use of motors with different duty cycles. NEC 620.13 requires the motor's nameplate current[6]** to be used when sizing feeder conductors. When the nameplate current is not available use the applicable full-load current listed in the Tables of Article 430. Depending upon the motor's duty cycle and time(duty) rating the nameplate current can be either increased or decreased per Table 430.22(E)[6]*** in accordance with NEC 620.13(A) and (D).

Continuous Duty Cycle

(1) Escalators and Moving Walks <NEC References - 620.13(A) & (D), 430.22(A) & (E) and 620.61(B)(2)>

List each escalator and moving walks motor. Use the given formula to perform load calculation.

Motor Load   Calculation (Refer to **Formula 6B**)

_____   _____
_____   _____
_____   _____

TOTAL = _____ VA
                (6F) (Combined)

Intermittent Duty Cycle

(1) Dumbwaiters <NEC References - 620.13(A) & (D), 430.22(A) & (E) and 620.61(B)(1)>

List each dumbwaiter motor load. Use the given formula to perform load calculation.

Motor Load   Calculation (Refer to **Formula 6B**)

_____   _____
_____   _____
_____   _____

TOTAL = _____ VA
                (6G) (Combined)

(2) Elevators <NEC References - 620.14, 620.13(A) & (D), 430.22(A) & (E) and 620.61(B)(1)>

List each elevator motor load. Use the given formula to perform load calculation.

Motor Load   Calculation (Refer to **Formula 6B**)

_____   _____
_____   _____
_____   _____

TOTAL = _____ VA
                (Combined)

Elevator's Demand Factor <NEC Reference - [6]****Table 620.14> and 430.22(E)

_____
(number [No.] of elevators)

_____ VA x _____ = _____ VA
(Combined)   (Demand Factor per No. of elevator motors)[6]****   **(6H)** (Demand Load)

(3) Stairway and Platform Lifts <NEC References - 620.13(A) & (D), 430.22(A) & (E) and 620.61(B)(4)>

List each stairway and platform lift motor. Use the given formula to perform load calculation. If VA value given computation not required.

Motor Load   Calculation (Refer to **Formula 6B**)

_____   _____
_____   _____
_____   _____

TOTAL = _____VA
                  (6I) (Combined)

E. Total Motor Loads - [LINE LOAD - Add lines **(6A)** - **(6I)**] - [NEUTRAL LOAD - Total motor load with neutral connections (120V)]

6.   <u>LINE LOAD</u>   [FN4] <u>NEUTRAL LOAD</u>
                                              [FN5] Permitted   Prohibited
                                                    Reduction   Reduction

       _____ VA   _____ VA   _____ VA

7. MEDICAL EQUIPMENT

A. Diagnostic Equipment

(1) Branch Circuits <NEC Reference - 517.73(A)(1)> (Individual branch circuit load [supplied separately] based on either the momentary or long-time rating of the given load. Momentary Rating [at 50%] - Long-Time Rating [at 100%]. Use the greater of the two ratings per circuit.)

List each piece of diagnostic equipment per voltage rating and use the given formulas to perform load calculation.

Branch Circuit   Calculation (Refer to **Formula 7A1**)

_____   _____
_____   _____
_____   _____

TOTAL = _____VA
                  (7A) (Combined)

Formula 7A1

(a) Momentary Rating [50% (.50)]

$$[(\underline{\qquad}V \times \underline{\qquad}A \times \underline{\qquad} \quad or \quad \underline{\qquad}VA)] \times .50 = \underline{\qquad}VA$$
$$\text{(Volts)} \qquad \text{(Amps)} \qquad \text{(If 3}\phi, \sqrt{3}) \qquad \text{(Given)}$$

(b) Long-Time Rating [100% (1)]

$$[(\underline{\qquad}V \times \underline{\qquad}A \times \underline{\qquad} \quad or \quad \underline{\qquad}VA)] = \underline{\qquad}VA$$
$$\text{(Volts)} \qquad \text{(Amps)} \qquad \text{(If 3}\phi, \sqrt{3}) \qquad \text{(Given)}$$

(2) Feeders <NEC Reference - 517.73(A)(2)> (Two or more branch circuits supplied from a common or separate feeders. Where simultaneous biplane examinations [SBE] are undertaken with X-ray units, 100% of the momentary demand rating of each X-ray unit is required.)

Feeder   Calculation (Refer to **Formula 7A2**)

_____   _____

_____   _____

_____   _____

TOTAL = _____VA
(7B) (Combined)

Formula 7A2

(a) Largest Unit - (50% [.50] of momentary demand rating)

$$[(\underline{\qquad}V \times \underline{\qquad}A \times \underline{\qquad} \quad or \quad \underline{\qquad}VA)] \times .50 = \underline{\qquad}VA$$
$$\text{(Volts)} \qquad \text{(Amps)} \qquad \text{(If 3}\phi, \sqrt{3}) \qquad \text{(Given)}$$

(b) Next Largest Unit - (25% [.25] of momentary demand rating)

$$[(\underline{\qquad}V \times \underline{\qquad}A \times \underline{\qquad} \quad or \quad \underline{\qquad}VA)] \times .25 = \underline{\qquad}VA$$
$$\text{(Volts)} \qquad \text{(Amps)} \qquad \text{(If 3}\phi, \sqrt{3}) \qquad \text{(Given)}$$

(c) Additional Units - (10% [.10] of momentary demand rating) (apply formula as often as needed)

$$[(\underline{\qquad}V \times \underline{\qquad}A \times \underline{\qquad} \quad or \quad \underline{\qquad}VA)] \times .10 = \underline{\qquad}VA$$
$$\text{(Volts)} \qquad \text{(Amps)} \qquad \text{(If 3}\phi, \sqrt{3}) \qquad \text{(Given)}$$

- or If -

(d) Simultaneous biplane examinations - (100% [1] of momentary demand rating)

$$[(\underline{\qquad}V \times \underline{\qquad}A \times \underline{\qquad} \quad or \quad \underline{\qquad}VA)] = \underline{\qquad}VA$$
$$\text{(Volts)} \qquad \text{(Amps)} \qquad \text{(If 3}\phi, \sqrt{3}) \qquad \text{(Given)}$$

B. Therapeutic Equipment

Branch Circuits Load <NEC Reference - 517.73(B)(1)>

List each piece of therapeutic equipment per voltage rating and use the given formula to perform load calculation.

Branch Circuit   Calculation [Use **Formula 7A2**(d)]

_____   _____
_____   _____
_____   _____

TOTAL = _____VA
(7C) (Combined)

C. Medical Equipment Total Load
[LINE LOAD - Add lines **(7A)** - **(7C)**] - [NEUTRAL LOAD - only applies where applicable]

7. LINE LOAD   [FN4] NEUTRAL LOAD
                                     [FN5] Permitted   Prohibited
                                           Reduction   Reduction

_____ VA   _____ VA   _____ VA

8. INDUSTRIAL EQUIPMENT

A. Electric Welders (Article 630)

Arc Welders

(1) Individual Welders <NEC Reference - 630.11(A)> [Where provided use the I_{1eff} Value[8*] given on the nameplate of the welding equipment. Where I_{1eff} value is not given, use primary current value[8**] given on welder rating plate and apply the appropriate multiplication factor per Table 630.11(A)[8***] based on the duty cycle and operating conditions of the welder.]

List each individual arc welder as provided per operating conditions and apply the given formula to perform load calculation.

Arc Welder   Calculation (Refer to **Formula 8AW1**)

_____   _____
_____   _____
_____   _____

TOTAL = _____VA
(8A) (Combined)

Formula 8AW1

$$\underline{\hspace{2cm}}\underset{I_{1eff}\text{ value}^{8*}}{}A \; or \; (\underline{\hspace{1.5cm}}\underset{Amps^{8**}}{}A \; x \; \underline{\hspace{1.5cm}}\underset{(\%)^{8***}}{}) \; x \; \underline{\hspace{1.5cm}}\underset{(Volts)}{}V \; x \; \underline{\hspace{1.5cm}}\underset{(If\;3\phi,\;\sqrt{3})}{} = \underline{\hspace{1.5cm}}\underset{(Total)}{}VA$$

-or-

(if (W)VA is provided based upon the welder's rating plate primary current)

$$\underline{\hspace{3cm}}(W)VA \; x \; \underline{\hspace{1.5cm}}\underset{(\%)^{8***}}{} = \underline{\hspace{1.5cm}}\underset{(Total)}{}VA$$

(2) Group of Welders <NEC Reference - 630.11(B)> [Where provided use the I_{1eff} Value^{8*} given on the nameplate of the welding equipment. Where I_{1eff} value is not given, use primary current value^{8**} given on welder rating plate and apply the appropriate multiplication factor per Table 630.11(A)^{8***} based on the duty cycle and operating conditions of the welder.]

List each individual arc welder as provided from largest to smallest current values based on the provisions of the given formula to perform load calculation.

Current Values (A)   Calculation (Refer to **Formula 8AW2**)

_____   _____
_____   _____
_____   _____
_____   _____

$$TOTAL = \underline{\hspace{2cm}}\underset{\textbf{(8B)}\;(Combined)}{}VA$$

Formula 8AW2

(a) Two Largest Welders [100% (1)]

$$\underline{\hspace{2cm}}\underset{I_{1eff}\text{ value}^{8*}}{}A \; or \; (\underline{\hspace{1.5cm}}\underset{Amps^{8**}}{}A \; x \; \underline{\hspace{1.5cm}}\underset{(\%)^{8***}}{}) \; x \; \underline{\hspace{1.5cm}}\underset{(Volts)}{}V \; x \; \underline{\hspace{1.5cm}}\underset{(If\;3\phi,\;\sqrt{3})}{} = \underline{\hspace{1.5cm}}\underset{(Total)}{}VA$$

$$\underline{\hspace{2cm}}\underset{I_{1eff}\text{ value}^{8*}}{}A \; or \; (\underline{\hspace{1.5cm}}\underset{Amps^{8**}}{}A \; x \; \underline{\hspace{1.5cm}}\underset{(\%)^{8***}}{}) \; x \; \underline{\hspace{1.5cm}}\underset{(Volts)}{}V \; x \; \underline{\hspace{1.5cm}}\underset{(If\;3\phi,\;\sqrt{3})}{} = \underline{\hspace{1.5cm}}\underset{(Total)}{}VA$$

(b) Third Largest Welder [85% (.85)]

$$\underline{\hspace{2cm}}\underset{I_{1eff}\text{ value}^{8*}}{}A \; or \; (\underline{\hspace{1.5cm}}\underset{Amps^{8**}}{}A \; x \; \underline{\hspace{1.5cm}}\underset{(\%)^{8***}}{}) \; x \; \underline{\hspace{1.5cm}}\underset{(Volts)}{}V \; x \; \underline{\hspace{1.5cm}}\underset{(If\;3\phi,\;\sqrt{3})}{} \; x \; .85 = \underline{\hspace{1.5cm}}\underset{(Total)}{}VA$$

(c) Fourth Largest Welder [70% (.70)]

$$\underline{\hspace{2cm}}\underset{I_{1eff}\text{ value}^{8*}}{}A \; or \; (\underline{\hspace{1.5cm}}\underset{Amps^{8**}}{}A \; x \; \underline{\hspace{1.5cm}}\underset{(\%)^{8***}}{}) \; x \; \underline{\hspace{1.5cm}}\underset{(Volts)}{}V \; x \; \underline{\hspace{1.5cm}}\underset{(If\;3\phi,\;\sqrt{3})}{} \; x \; .70 = \underline{\hspace{1.5cm}}\underset{(Total)}{}VA$$

(d) Remaining Welders [60% (.60)] (apply formula as often as needed)

$$\underset{I_{1eff}\text{ value}^{8*}}{\underline{\hspace{2cm}}}A \ or \ (\underset{\text{Amps}^{8**}}{\underline{\hspace{2cm}}}A \ x \underset{(\%)^{8***}}{\underline{\hspace{1.5cm}}}) \ x \underset{\text{(Volts)}}{\underline{\hspace{1.5cm}}}V \ x \underset{\text{(If 3}\phi, \sqrt{3})}{\underline{\hspace{1.5cm}}} x \ .60 = \underset{\text{(Total)}}{\underline{\hspace{1.5cm}}}VA$$

Resistance Welders

(1) Individual Welders (Use the primary current value[8•] given on welder rating plate.)

List each individual resistance welder as provided per operating conditions and apply the given formula to perform load calculation.

Resistance Welder   Calculation (Refer to **Formula 8RW1**)

_____   _____

_____   _____

_____   _____

$$\text{TOTAL} = \underset{\textbf{(8C)}\text{ (Combined)}}{\underline{\hspace{3cm}}}VA$$

Formula 8RW1

(a) Seam and Automatically fed <NEC Reference - 630.31(A)(1)>

$$(\underset{\text{(Volts)}}{\underline{\hspace{1cm}}}V \ x \underset{\text{(Amps)}^{8\bullet}}{\underline{\hspace{2cm}}}A) \ or \underset{\text{(Given)}}{\underline{\hspace{2cm}}}VA \ x \ .70 \ (70\%) = \underset{\text{(Total)}}{\underline{\hspace{2cm}}}VA$$

(b) Manually Operated Nonautomatic <NEC Reference - 630.31(A)(1)>

$$(\underset{\text{(Volts)}}{\underline{\hspace{1cm}}}V \ x \underset{\text{(Amps)}^{8\bullet}}{\underline{\hspace{2cm}}}A) \ or \underset{\text{(Given)}}{\underline{\hspace{2cm}}}VA \ x \ .50 \ (50\%) = \underset{\text{(Total)}}{\underline{\hspace{2cm}}}VA$$

(c) Actual Primary Current and Duty Cycle known (remains unchanged) <NEC Reference - 630.31(A)(2)> (Apply Table 630.31(A)(2) [8••] per welder's duty cycle).

$$\underset{\text{Amps}^{8\bullet}}{\underline{\hspace{2cm}}}A \ x \underset{(\%)^{8\bullet\bullet}}{\underline{\hspace{2cm}}} x \underset{\text{(Volts)}}{\underline{\hspace{1.5cm}}}V \ x \underset{\text{(If 3}\phi, \sqrt{3})}{\underline{\hspace{2cm}}} = \underset{\text{(Total)}}{\underline{\hspace{2cm}}}VA$$

or

$$\underset{\text{(Given)}}{\underline{\hspace{2cm}}}VA \ x \underset{(\%)^{8\bullet\bullet}}{\underline{\hspace{2cm}}} = \underset{\text{(Total)}}{\underline{\hspace{2cm}}}VA$$

(2) Group of Welders <NEC Reference - 630.31(B)> (Where two or more resistance welders are used, the largest welder and other welder's loads are combined.)

— The 2011 National Electrical Code Book of In-Depth Calculations — Volume 4 —

List the largest welder and the other (remaining) welders per primary current[8•] applying the appropriate multiplier per duty cycle as given in Table 630.31(A)(2)[8••] Use the given formula to perform load calculations.

Resistance Welder   Calculation (Refer to **Formula 8RW2**)

_____

_____

_____

                        TOTAL = _____ VA
                                    (8D) (Combined)

Formula 8RW2

(a) Largest Welder [Per 630.31(A)]

$$\underset{\text{Amps}^{8\bullet}}{\_\_\_\_} \text{A} \ \text{x} \ \underset{(\%)^{8\bullet\bullet}}{\_\_\_\_} \text{x} \ \underset{\text{(Volts)}}{\_\_\_\_} \text{V} \ \text{x} \ \underset{\text{(If 3}\phi, \sqrt{3})}{\_\_\_\_} = \underset{\text{(Total)}}{\_\_\_\_} \text{VA}$$

(b) Remaining Welders [60% (.60)] (apply formula as often as needed)

$$\underset{\text{Amps}^{8\bullet}}{\_\_\_\_} \text{A} \ \text{x} \ \underset{(\%)^{8\bullet\bullet}}{\_\_\_\_} \text{x} \ \underset{\text{(Volts)}}{\_\_\_\_} \text{V} \ \text{x} \ \underset{\text{(If 3}\phi, \sqrt{3})}{\_\_\_\_} \text{x} .60 = \underset{\text{(Total)}}{\_\_\_\_} \text{VA}$$

B. X-Ray Equipment (Article 660) - (For industrial or other nonmedical or nondental use.)

(1) Branch-Circuit Conductors <NEC Reference - 660.6(A)> (Individual branch circuit load [supplied separately] based on either the momentary or long-time rating of the given load. Momentary Rating [at 50%] - Long-Time Rating [at 100%]. Use the greater of the two ratings per circuit.)

List each piece of x-ray equipment per voltage rating and use the given formulas to perform load calculation.

Branch Circuit   Calculation (Refer to **Formula 8XRE1**)

_____

_____

                        TOTAL = _____ VA
                                  (8E) (Combined)

Formula 8XRE1

(a) Momentary Rating [50% (.50)]

$$[(\underset{\text{(Volts)}}{\_\_\_}\text{V} \ \text{x} \ \underset{\text{(Amps)}}{\_\_\_}\text{A} \ \text{x} \ \underset{\text{(If 3}\phi, \sqrt{3})}{\_\_\_} \ \text{or} \ \underset{\text{(Given)}}{\_\_\_}\text{VA})] \ \text{x} \ .50 = \underset{}{\_\_\_} \text{VA}$$

(b) Long-Time Rating [100% (1)]

$$[(\underline{\quad}\underset{\text{(Volts)}}{V} \times \underline{\quad}\underset{\text{(Amps)}}{A} \times \underline{\quad}\underset{\text{(If 3}\phi, \sqrt{3})}{} \text{ or }\underline{\quad}\underset{\text{(Given)}}{VA})] = \underline{\quad}VA$$

(2) Feeder Conductors <NEC Reference - 660.6(B)> (Applies when feeders will supply two or more x-ray equipment branch circuits.

Feeder/Service Calculation (Refer to **Formula 8XRE2**)

_____ _____

_____ _____

$$\text{TOTAL} = \underline{\qquad}VA$$
$$\underset{\textbf{(8F)}\text{ (Combined)}}{}$$

Formula 8XRE2

(a) Two Largest X-Ray Apparatus - (100% of momentary demand rating) as determined by 660.6(A).

$$[(\underline{\quad}\underset{\text{(Volts)}}{V} \times \underline{\quad}\underset{\text{(Amps)}}{A} \times \underline{\quad}\underset{\text{(If 3}\phi, \sqrt{3})}{} \text{ or }\underline{\quad}\underset{\text{(Given)}}{VA})] = \underline{\quad}VA$$

$$[(\underline{\quad}\underset{\text{(Volts)}}{V} \times \underline{\quad}\underset{\text{(Amps)}}{A} \times \underline{\quad}\underset{\text{(If 3}\phi, \sqrt{3})}{} \text{ or }\underline{\quad}\underset{\text{(Given)}}{VA})] = \underline{\quad}VA$$

(b) Other X-Ray Apparatus [Momentary Rating (at 20% (.20)] (apply formula as often as needed)

$$[(\underline{\quad}\underset{\text{(Volts)}}{V} \times \underline{\quad}\underset{\text{(Amps)}}{A} \times \underline{\quad}\underset{\text{(If 3}\phi, \sqrt{3})}{} \text{ or }\underline{\quad}\underset{\text{(Given)}}{VA})] \times .20 = \underline{\quad}VA$$

C. Total Special Equipment Loads
[LINE LOAD - Add lines **(8A)** - **(8F)**] – [NEUTRAL LOAD - Loads with neutral connections]

8. LINE LOAD

[FN4] NEUTRAL LOAD
[FN5] Permitted   Prohibited
    Reduction   Reduction

_____ VA   _____ VA   _____ VA

9. HEATING and AIR-CONDITIONING (AC) EQUIPMENT <NEC References - 220.50, 220.51[9]*, 220.60, 430.6(A)(1), and 440.6(A)> [9]**(If air-conditioning equipment consist of a hermetic [sealed] motor-compressor, use **(a)** the branch-circuit selection current [bcsc] [NEC 440.6(A), *Exception No. 1*] marked on the equipment. If equipment not marked with bcsc, use **(b)** rated load current(s) marked on the nameplate of the equipment or when no rated-load current(s) is shown on the equipment nameplate, use **(c)** the rated-load current on the compressor nameplate [NEC 440.6(A)]. When a conventional motor is used to drive an air-conditioning compressor use the full-load current [FLC] per motor's horsepower rating given in the FLC Tables of Article 430. For other type motors affiliated with either the Heating or AC equipment use the nameplate current marked on the equipment.

Note - Where VA loads are given computation is not required. If heat pump, add compressor load and the maximum amount of electric heat that will be energized simultaneously [at the same time].) Apply, if Heat Pump is used entirely for heating and air-conditioning load. If not, use larger of Heating and Air-Conditioning loads. Larger load to include all loads that could operate at the same time, *example*: where Heat Pump is used for supplemental needs add to larger load if applicable. Use the following formulas to perform required load calculations.

ELECTRICAL HEATING (EH) UNITS

(1) Electric Heat - _____ W (VA) x _____ x _____ = _____ VA
  (Unit Rating)[+]   (Percentage)* (No.)***   (Total)

(2) Electric Heat - _____ W (VA) x _____ x _____ = _____ VA
  (Unit Rating)[+]   (Percentage)* (No.)***   (Total)

(3) Electric Heat - _____ W (VA) x _____ x _____ = _____ VA
  (Unit Rating)[+]   (Percentage)* (No.)***   (Total)

[+]If calculation required, use the following formula.   TOTAL HEAT = _____ VA
  (V x A) x ($\sqrt{3}$-if applicable) = Unit Rating   (Combined)

AIR-CONDITIONING (AC) UNITS

(1) AC - _____ VA[a] + _____ VA[b] x _____ x _____ = _____ VA
  (Voltage x amps**)   (Voltage x amps**)   (if 3φ, x $\sqrt{3}$) (No.)*** (Total)

(2) AC - _____ VA[a] + _____ VA[b] x _____ x _____ = _____ VA
  (Voltage x amps**)   (Voltage x amps**)   (if 3φ, x $\sqrt{3}$) (No.)*** (Total)

(3) AC - _____ VA[a] + _____ VA[b] x _____ x _____ = _____ VA
  (Voltage x amps**)   (Voltage x amps**)   (if 3φ, x $\sqrt{3}$) (No.)*** (Total)

                                        TOTAL AC = _____ VA
                                                      (Combined)

HEAT PUMP (HTP) UNITS

Heat Pump Load

(1) Heat Pump - _____ VA[a] + _____ VA[b] +
  (Voltage x amps**)   (Voltage x amps**)

  _____ W(VA) x _____ x _____ = _____ VA
  (Maximum Electric Heat)   (if 3φ, x $\sqrt{3}$) (No.)*** (Total)

(2) Heat Pump - _____ VA^a + _____ VA^b +
(Voltage x amps**)   (Voltage x amps**)

_____ W(VA) x _____ x _____ = _____ VA
(Maximum Electric Heat)   (If 3φ, √3)   (No.)***   (Total)

(3) Heat Pump - _____ VA^a + _____ VA^b +
(Voltage x amps**)   (Voltage x amps**)

_____ W(VA) x _____ x _____ = _____ VA
(Maximum Electric Heat)   (If 3φ, √3)   (No.)***   (Total)

TOTAL HEAT PUMP = _____ VA
(Combined)

^{a}Compressor ^{b}Fan Motor
**RLA (running load amps) or FLC (full-load current) ***Number of units with identical operating characteristics.

LINE LOAD = _____ VA + _____ VA + _____ VA
(Larger Load)   (Other[s])•   (Total)

•Blowers, Air Handlers, etc. – if applicable. (Use [V x A] if 3φ, x √3 when VA not given).

9.   LINE LOAD   [FN4] NEUTRAL LOAD
                                          [FN5] Permitted   Prohibited
                                               Reduction   Reduction

_____ VA   _____ VA _____ VA

10. LARGEST MOTOR <NEC References - 430.17, 430.24, 440.7 and 440.33> (Use motor with highest current per NEC 430.17 and 440.7)

_____ V x _____ A x _____ = _____ VA x .25 (25%) = _____ VA
(Volts)   (FLC/RLA)   (If 3φ, √3)   (Computed/Given)

10.   LINE LOAD   [FN4] NEUTRAL LOAD
                                          [FN5] Permitted   Prohibited
                                               Reduction   Reduction

_____ VA   _____ VA _____ VA

TOTAL DEMAND LOAD (LINE and NEUTRAL) (List each computed line and neutral loads below and total lines 1. – 10.)

	LINE LOAD	NEUTRAL LOAD	
		Permitted Reduction	Prohibited Reduction
1. General Lighting	_____ VA	_____ VA	_____ VA
2. Other Lighting Loads	_____ VA	_____ VA	_____ VA
3. Receptacle Loads	_____ VA	_____ VA	_____ VA
4. Kitchen Equipment	_____ VA	_____ VA	_____ VA
5. Specific Loads	_____ VA	_____ VA	_____ VA
6. Motor Loads	_____ VA	_____ VA	_____ VA
7. Medical Equipment	_____ VA	_____ VA	_____ VA
8. Industrial Equipment	_____ VA	_____ VA	_____ VA
9. Heating and AC Equip.	_____ VA	_____ VA	_____ VA
10. Largest Motor	_____ VA	_____ VA	_____ VA
Total Demand Load (VA) =	_____ VA	_____ VA	_____ VA

OCCUPANCY'S OPERATING LINE VOLTAGE - _____ V _____ V
(Given operating voltage or as determined by other)   (1ϕ)   (3ϕ)

CALCULATE MINIMUM LINE and NEUTRAL LOADS
(Divide Total Demand Load [VA] by operating line voltage [V])

LINE LOAD   = _____ VA / (_____ V [x $\sqrt{3}$, If 3ϕ]) = _____ A

NEUTRAL LOAD

   Permitted   = _____ VA / (_____ V [x $\sqrt{3}$, If 3ϕ]) = _____ A*

   Prohibited   = _____ VA / (_____ V [x $\sqrt{3}$, If 3ϕ]) = _____ A

                          NEUTRAL LOAD = _____ A**

*Where the feeder or service neutral load exceeds 200A, NEC 220.61(B)(2) permits the load to be reduced by 70 percent. However, this reduction is not permitted when the feeder or service neutral load consist of nonlinear loads (refer to the description of footnote 5 [**FN5**]). Complete the following to determine the permitted Neutral Load.

_____ A – 200A = _____ x .70 = _____ A
(Permitted Reduction Load)   (Remainder)   (Permitted Reduction)

_____ A + 200A = _____ A
(Permitted Reduction)   (Permitted Neutral Load)

**FEEDER or SERVICE NEUTRAL LOAD = Permitted Load + Prohibited Load

SIZE SERVICE (Size of service based on the calculated **LINE LOAD**)

SIZE SERVICE REQUIRED (minimum) _____A

SIZING FEEDER/SERVICE CONDUCTORS - (Based on the calculated LINE LOAD and NEC References.) <NEC References - 215.2(A), 230.42(A), 240.4(B) & (C), 310.10(H), 310.15(B)(2) & (3), and Table 310.15(B)(16)> (Based on the calculated LINE LOAD and NEC References.)

FEEDER/SERVICE CONDUCTORS _____

SIZING NEUTRAL CONDUCTOR <NEC References - 215.2(A)(2), 220.61, 230.42(C), 250.24(C), 310.10(H), 310.15(B)(2), (3) & (5) and Table 310.15(B)(16)> (Based on the calculated NEUTRAL LOAD and NEC References.)

NEUTRAL CONDUCTOR(S) _____

SIZING GROUNDING ELECTRODE CONDUCTOR <NEC References - 250.24(C), 250.66 & Table 250.66 and Table 8 of Chapter 9>

GROUNDING ELECTRODE CONDUCTOR _____

STANDARD LOAD CALCULATIONS for HOTELS and MOTELS
---

1. GENERAL LIGHTING and RECEPTACLE LOADS <NEC References - 220.12, Table 220.12, 220.14(J), 220.42 and Table 220.42> Where actual lighting loads are applied receptacle loads must be calculated separately. Where receptacle loads are calculated separately or considered continuous refer to Item **3**. - Receptacles Load.

FN = Footnote $\sqrt{3} = 1.732$

General Lighting Load* or **General Lighting and Receptacle Loads** (Square Feet [SF])
*when receptacle load calculated separately

_____ SF x _____ x 2VA = _____ VA
(Unit dimensions 1)   (No. of units)

_____ SF x _____ x 2VA = _____ VA
(Unit dimensions 2)   (No. of units)

_____ SF x _____ x 2VA = _____ VA
(Unit dimensions 3)   (No. of units)

_____ SF x _____ x 2VA = _____ VA
(Unit dimensions 4)   (No. of units)

    A. General Lighting Load or **General
      Lighting** and **Receptacle Loads**   = _____ VA

Actual Lighting Load (List all fixtures that will contribute to the occupancy's entire interior lighting load)

Type Fixture	VA rating [FN1]	No. of Fixtures		TOTAL VA
_____	_____ x	_____	=	_____
_____	_____ x	_____	=	_____
_____	_____ x	_____	=	_____
_____	_____ x	_____	=	_____
_____	_____ x	_____	=	_____

    B. Actual Lighting Load   =   _____ VA
                                                 (or given Load)

[FN1] - If INCANDESCENT FIXTURE, use wattage of lamp (bulb). For lighting units that have BALLAST, TRANSFORMERS, AUTOTRANSFORMERS, or LED DRIVERS, the calculated load shall be based on the total ampere ratings of such units [NEC 220.18(B)].

Apply the **LARGER VA** of **A.** and **B.**

LARGER = _____ VA

APPLY DEMAND FACTORS (if applicable) <NEC References - 220.42 and Table 220.42 (Includes apartment houses without provisions for cooking by tenants.)> If LARGER VA less than or equal to 100,000VA, step **c.** is not required.

a. First 20,000VA or less of LARGER VA (at 50%)

_____ x .50   = _____ VA

b. _____ x .40 (at 40%)   = _____ VA
    (LARGER VA – 20,001VA up to 100,000VA)

c. _____ x .30 (at 30%)   = _____ VA
    (Remainder of LARGER VA exceeding 100,000VA)*

    *(LARGER VA – 100,000 VA = _____ VA)

TOTAL (Lines **a.** - **c.**)   = _____ VA
(Derated General Lighting and Receptacle Loads)   (Enter value below -
                                                        LINE and NEUTRAL)

GENERAL LIGHTING and RECEPTACLE LOADS

1. LINE LOAD   [FN2] NEUTRAL LOAD
                                [FN3] Permitted   Prohibited
                                      Reduction   Reduction

_____ VA   _____ VA   _____ VA

[FN2] - **Neutral loads** are computed at 100 percent. Although the neutral conductor(s) will serve continuous loads, under most conditions it will never experience the same demands as the line conductors and is subject to derating per NEC 220.61. For clarity, Neutral Loads as it pertains to this worksheet are recognized as those loads that are associated with either **(1)** a *neutral conductor* as described in Article 100 *or* **(2)** a *grounded conductor* which carries the same amount of current as an ungrounded conductor *or* **(3)** a *common conductor* as referenced in NEC 310.15(B)(5). See Article 310 (NEC 310.15(B)(5) [Volume 2].

[FN3] - Reduction of the feeder or service neutral load is **permitted** to have an additional 70 percent (.70) applied when supplying household electric ranges, wall-mounted ovens, counter-mounted cooking units and dryers. This 70 percent (.70) reduction is also permitted for unbalanced loads in excess of 200 amperes where the feeder or service supply *linear loads** from a 3-wire dc *or* single-phase ac system, *or* a 4-wire, 3-phase; 3-wire, 2-phase-system, *or* a 5-wire, 2-phase system.

Reduction of the feeder or service neutral load is **prohibited** for that portion of the load that consist of: **(1)** a 3-wire circuit consisting of 2 ungrounded conductors (208/120V- 1φ) and the neutral of a 4-wire, 3-phase, wye-connected system (208/120V-3φ) *and* **(2)** that portion of the load consisting of *nonlinear loads*** supplied from a 4-wire, wye connected, 3-phase system (208/120V- 3φ and 480/277V- 3φ).

*Examples of *linear loads*: Heating equipment, electric motors, resistive lighting (incandescent), etc.
**Examples of *nonlinear loads*: computer equipment, converters, data-processing equipment, drives (adjustable/frequency/speed/variable), electronic ballast, electric discharge lighting (fluorescent, high and low-pressure sodium, mercury-vapor, metal-halide, etc.), inverters, medical and laboratory test equipment, programmable logic controllers (PLC), UPS systems, welders, etc.

2. OTHER LIGHTING LOADS

A. Sign/Outline (S/O) Lighting <NEC References - 220.12(F) and 600.5(A)>

S/O 1 - _____ V x _____ A (or given VA(W) Load) = _____ VA
        (Voltage)   (Amperes)   (Minimum of 1200VA required)

S/O 2 - _____ V x _____ A (or given VA(W) Load) = _____ VA
        (Voltage)   (Amperes)   (Minimum of 1200VA required)

S/O 3 - _____ V x _____ A (or given VA(W) Load) = _____ VA
        (Voltage)   (Amperes)   (Minimum of 1200VA required)

Total (Sign/Outline Lighting) = _____ VA **(2A)**

B. Outside Lighting <NEC Reference - 220.18(B)>

_____ ____V x ____ A or _____W(VA) x _____[FN4] = _____ VA
(Type - 1) (Volts) (Amps) (Lamp Watts) (No.) (Load)•

_____ ____V x ____ A or _____W(VA) x _____[FN4] = _____ VA
(Type - 2) (Volts) (Amps) (Lamp Watts) (No.) (Load)•

_____ ____V x ____ A or _____W(VA) x _____[FN4] = _____ VA
(Type - 3) (Volts) (Amps) (Lamp Watts) (No.) (Load)•

[FN4] - Number of Fixtures with identical operating characteristics. •Calculated or provided by other.

Total (Outside Lighting) = _____ VA **(2B)**

C. Show-Window Lighting <NEC Reference - 220.43(A)> (Voltage rating _____V)

_____ x 200VA (or given VA Load) = _____ VA
(Linear Feet)   **(2C)**

D. Track Lighting <NEC Reference - 220.43(B)> (Voltage rating _____V)

(_____ ÷ 2') x 150VA (or given VA Load) = _____ VA
(Linear Feet)   **(2D)**

E. Miscellaneous (Write-ins. List individual voltage rating of each lighting [fixture] load.)

_____

_____

_____

Total (Miscellaneous) = _____ VA
                              (2E)

F. Other Lighting (LINE and NEUTRAL) Loads Total [Add lines **(2A)** - **(2E)**, where applicable]

TOTAL = _____ VA x 1.25 [FN5] + _____ = _____ VA
         (Continuous)   (Noncontinuous)   (LINE LOAD)

[FN5] - Percentage increase only apply to LINE LOAD – NEUTRAL LOAD calculated at 100 percent.

2. <u>LINE LOAD</u>   [FN2] <u>NEUTRAL LOAD</u>
                                             (120V or 277V)
                                    [FN3] Permitted   Prohibited
                                          Reduction   Reduction

    _____ VA   _____ VA   _____ VA

3. RECEPTACLE LOADS (120 volts only)

A. Non-continuous duty [FN6] <NEC References 220.14(H) and (I), 220.44 and Table 220.44>

   (1) Fixed Multioutlet Assemblies [FN7] <220.14(H)>

   Non-simultaneous use
   (_____ ÷ 5) x 180VA = _____ VA
    (Linear Feet)   **(3A)**

   Simultaneous use
   _____ x 180VA = _____ VA
    (Linear Feet)   **(3A)**

   [FN7] - Usually not considered a continuous load. However if so, include that portion with **3E** and proceed.

   (2) General Purpose Receptacles* and Fixed Multioutlet Assemblies <220.14(I)>

   _____ x 180VA = _____ VA + _____ VA = _____ VA
   (No. of receptacles)*   **(3A)** + **(3B)**   (TOTAL)

   *Only apply if other than bank or office building. For bank and office buildings apply item **B**.

— 64 —

[FN6] - APPLY DEMAND FACTORS [If applicable]

a. First 10,000VA (10kVA) of above TOTAL   = _____ VA
(At 100 percent)

b. _____ x .50   = _____ VA
(Remainder of TOTAL exceeding 10,000VA)**

**(_____ VA (TOTAL) – 10,000 VA = _____ VA)

_____ VA
(3C) [Total **(2)a.** and **b.**]

B. Non-continuous duty (If Bank or Office Building [FN8]) <NEC Reference - 220.14(K)>

[FN8] General Purpose Receptacles [Use Larger of **(1)** or **(2)** – Because item **(1)** is derived from NEC 220.14(I), the provisions of NEC and Table 220.44 are permitted]

(1) _____ x 180VA·   = _____ VA
(No. of receptacles)

[FN6] - APPLY DEMAND FACTORS [if applicable]. If the results of **(1)** is larger than **(2)**, omit **B.** and include the number of receptacles with step **3.A.(2)** if applicable. Apply demand factors if the number of receptacles and step **3.A.(1)** exceeds 10,000VA.

a. First 10,000VA (10kVA) of above TOTAL   = _____ VA
(At 100 percent)

b. _____ x .50   = _____ VA
(Remainder of TOTAL exceeding 10,000VA)**

**(_____ VA (TOTAL) – 10,000 VA = _____ VA)

(2) _____ x 1 VA/SF = _____ VA
Building's Square Footage

_____ VA
(3D) [Larger of Total - **(1)a.** and **b.** or **(2)**]

C. Continuous duty <NEC References - 220.14(I), 215.2(A)(1) [FEEDER] or 230.42(A) [SERVICE]

_____ x 180VA· (or 90VA··) = _____ VA x 1.25 = _____ VA
(No. of receptacles)   **(3E)**   **(3F)**

·For each single/multiple receptacle on one strap. Single piece comprised of 4 or more receptacles, 90VA.

D. Total Receptacle (LINE and NEUTRAL) Load
[LINE LOAD - Add lines **(3C)**, **(3D)** and **(3F)**]   [NEUTRAL LOAD - Add lines **(3C)**, **(3D)** and **(3E)**]

3.   <u>LINE LOAD</u>   [FN2] <u>NEUTRAL LOAD</u>

                                                         [FN3] Permitted   Prohibited

                                                             Reduction   Reduction

_____ VA   _____ VA _____ VA

4. OTHER LOADS (Continuous and Noncontinuous)

List continuous and noncontinuous loads and calculate volt-ampere (VA) ratings if not given. If 3-phase, apply √3 (1.732).

<u>Continuous Loads</u>   <u>Calculations</u>

_____   _____

_____   _____

_____   _____

_____   _____

_____   _____

                                         TOTAL = _____
                                                     (4A)

<u>Noncontinuous Loads</u>   <u>Calculations</u>

_____   _____

_____   _____

_____   _____

_____   _____

_____   _____

                                         TOTAL = _____
                                                     (4B)

Total Other Loads - _____ VA x 1.25[FN5] + _____ VA = _____ VA
                       (4A)   (4B)   Total

[FN5] - Percentage increase only apply to LINE LOAD – NEUTRAL LOAD calculated at 100 percent.

4.   <u>LINE LOAD</u>   [FN2] <u>NEUTRAL LOAD</u>

                                                          (120V or 277V)

                                                       [FN3] Permitted   Prohibited

                                                              Reduction   Reduction

_____ VA   _____ VA _____ VA

5. HEATING and AIR-CONDITIONING (AC) EQUIPMENT <NEC References - 220.50, 220.51*, 220.60, 430.6(A)(1), and 440.6(A)> **(If air-conditioning equipment consist of a hermetic [sealed] motor-compressor, use (a) the branch-circuit selection current [bcsc] [NEC 440.6(A), *Exception No. 1*] marked on the equipment. If equipment not marked with bcsc, use

(b) rated load current(s) marked on the nameplate of the equipment or when no rated-load current(s) is shown on the equipment nameplate, use **(c)** the rated-load current on the compressor nameplate [NEC 440.6(A)]. When a conventional motor is used to drive an air-conditioning compressor use the full-load current [FLC] per motor's horsepower rating given in the FLC Tables of Article 430. For other type motors affiliated with either the Heating or AC equipment use the nameplate current marked on the equipment.

Note - Where VA loads are given computation is not required. If heat pump, add compressor load and the maximum amount of electric heat that will be energized simultaneously [at the same time].) Apply, if Heat Pump is used entirely for heating and air-conditioning load. If not, use larger of Heating and Air-Conditioning loads. Larger load to include all loads that could operate at the same time, *example*: where Heat Pump is used for supplemental needs add to larger load if applicable. Use the following formulas to perform required load calculations.

ELECTRICAL HEATING (EH) UNITS

(1) Electric Heat - _____ W (VA) x _____ x _____ = _____ VA
(Unit Rating)+ (Percentage)* (No.)*** (Total)

(2) Electric Heat - _____ W (VA) x _____ x _____ = _____ VA
(Unit Rating)+ (Percentage)* (No.)*** (Total)

(3) Electric Heat - _____ W (VA) x _____ x _____ = _____ VA
(Unit Rating)+ (Percentage)* (No.)*** (Total)

+If calculation required, use the following formula. TOTAL HEAT = _____ VA
(V x A) x (√3-if applicable) = Unit Rating   (Combined)

AIR-CONDITIONING (AC) UNITS

(1) AC - _____VA^a + _____VA^b x_____ x _____ = _____ VA
(Voltage x amps**) (Voltage x amps**) (if 3φ, x √3) (No.)*** (Total)

(2) AC - _____VA^a + _____VA^b x_____ x _____ = _____ VA
(Voltage x amps**) (Voltage x amps**) (if 3φ, x √3) (No.)*** (Total)

(3) AC - _____VA^a + _____VA^b x_____ x _____ = _____ VA
(Voltage x amps**) (Voltage x amps**) (if 3φ, x √3) (No.)*** (Total)

TOTAL AC = _____VA
(Combined)

HEAT PUMP (HTP) UNITS

Heat Pump Load

(1) Heat Pump - _____ VA^a + _____ VA^b +
(Voltage x amps**) (Voltage x amps**)

_____ W(VA) x _____ x _____ = _____ VA
(Maximum Electric Heat) (if 3φ, x √3) (No.)*** (Total)

(2) Heat Pump - _____ VA^a + _____ VA^b +
(Voltage x amps**) (Voltage x amps**)

_____ W(VA) x _____ x _____ = _____ VA
(Maximum Electric Heat) (If 3φ, √3) (No.)*** (Total)

(3) Heat Pump - _____ VA^a + _____ VA^b +
(Voltage x amps**) (Voltage x amps**)

_____ W(VA) x _____ x _____ = _____ VA
(Maximum Electric Heat) (If 3φ, √3) (No.)*** (Total)

TOTAL HEAT PUMP = _____ VA
(Combined)

^{a}Compressor ^{b}Fan Motor
**RLA (running load amps) or FLC (full-load current) ***Number of units with identical operating characteristics.

LINE LOAD = _____ VA + _____ VA + _____ VA
(Larger Load) (Other[s])• (Total)

•Blowers, Air Handlers, etc. – if applicable. (Use [V x A] if 3φ, x √3 when VA not given).

5.   LINE LOAD   [FN2] NEUTRAL LOAD
                                         [FN3] Permitted   Prohibited
                                              Reduction   Reduction

_____ VA   _____ VA _____ VA

6. LARGEST MOTOR <NEC References - 430.17, 430.24, 440.7 and 440.33> (Use motor with highest current per NEC 430.17 and 440.7)

_____V x _____A x _____ = _____ VA x .25 (25%) = _____VA
(Volts) (FLC/RLA) (If 3φ, √3) (Computed/Given)

6.   LINE LOAD   [FN2] NEUTRAL LOAD
                                         [FN3] Permitted   Prohibited
                                              Reduction   Reduction

_____ VA   _____ VA _____ VA

TOTAL DEMAND LOAD (LINE and NEUTRAL) (List each computed line and neutral loads below and total lines 1. - 6.)

	LINE LOAD	NEUTRAL LOAD	
		Permitted Reduction	Prohibited Reduction
1. General Lighting	_____ VA	_____ VA	_____ VA
2. Other Lighting Loads	_____ VA	_____ VA	_____ VA
3. Receptacle Loads	_____ VA	_____ VA	_____ VA
4. Other Loads	_____ VA	_____ VA	_____ VA
5. Heating and AC Equip.	_____ VA	_____ VA	_____ VA
6. Largest Motor	_____ VA	_____ VA	_____ VA
Total Demand Load (VA) =	_____ VA	_____ VA	_____ VA

OCCUPANCY'S OPERATING LINE VOLTAGE -   _____ V   _____ V

(Given operating voltage or as determined per test examination)   (1ϕ)   (3ϕ)

CALCULATE MINIMUM LINE and NEUTRAL LOADS
(Divide Total Demand Load [VA] by operating line voltage [V])

LINE LOAD   =   _____ VA / (_____ V [x $\sqrt{3}$, If 3ϕ]) = _____ A

NEUTRAL LOAD

   Permitted   =   _____ VA / (_____ V [(x $\sqrt{3}$, If 3ϕ]) = _____ A*

   Prohibited   =   _____ VA / (_____ V [(x $\sqrt{3}$, If 3ϕ]) = _____ A

NEUTRAL LOAD = _____ A**

*Where the feeder or service neutral load exceeds 200A, NEC 220.61(B)(2) permits the load to be reduced by 70 percent. However, this reduction is not permitted when the feeder or service neutral load consist of nonlinear loads (refer to the description of footnote 3 [**FN3**]). Complete the following to determine the permitted Neutral Load.

_____ A − 200A = _____ x .70 = _____ A
(Permitted Reduction Load)   (Remainder)   (Permitted Reduction)

_____ A + 200A = _____ A
(Permitted Reduction)   (Permitted Neutral Load)

**FEEDER or SERVICE NEUTRAL LOAD = Permitted Load + Prohibited Load

SIZE SERVICE (Size of service based on the calculated **LINE LOAD**)

SIZE SERVICE REQUIRED (minimum) _____A

SIZING FEEDER/SERVICE CONDUCTORS <NEC References - NEC 215.2(A), 230.42(A), 240.4(B) & (C), 310.10(H), 310.15(B)(2) & (3) and Table 310.15(B)(16)> (Based on the calculated LINE LOAD and NEC References.)

FEEDER/SERVICE CONDUCTORS _____

SIZING NEUTRAL CONDUCTOR <NEC References - 215.2(A)(2), 220.61, 230.42(C), 250.24(C), 310.10(H), 310.15(B)(2), (3) & (5) and Table 310.15(B)(16)> (Based on the calculated NEUTRAL LOAD and NEC References.)

NEUTRAL CONDUCTOR(S) _____

SIZING GROUNDING ELECTRODE CONDUCTOR <NEC References - 250.24(C), 250.66 & Table 250.66 and Table 8 of Chapter 9>

GROUNDING ELECTRODE CONDUCTOR _____

ABOUT THE AUTHOR

For over thirteen years, Alvin Walker, a native of Shreveport, Louisiana owned and operated a small yet successful electrical contracting business. He now works as an author and instructor specializing in electrical and NEC training where his services are available throughout the United States. In his over thirty year of experience, he has developed a very strong background in electrical engineering, electrical design, electrical maintenance and construction. He has taught Business Law for Contractors, the National Electrical Code, Electrical Theory, and other basic and advanced electrical classes at Bossier Parish Community College, Louisiana State University-Shreveport, Northwest Louisiana Technical College (Forcht Wade Correction Center), and Southern University-Shreveport to include once serving as the Department Head of Industrial Electricity at Houston Community College-Stafford, Texas.

Mr. Walker is best known for his hands-on approach and the ability to simplify and explain the most difficult electrical subject matters. He is a master electrician and holds a Louisiana state license as an electrical contractor. He has a degree in electrical engineering from the University of South of Carolina and has worked as an electrical engineer for E.I. DuPont and Westinghouse at The Savannah River Plant (Company) of Aiken, South Carolina and M.W. Kellogg of Houston, Texas.

In his daily life Mr. Walker is a devoted Christian who has a passion for serving Christ, his fellowman and teaching and spreading the Word of God. As a recipient of three honorable discharges, he served over 9 years in the United States Army.

He enjoys traveling, wood-works and carpentry but is best known for his famous smoked barbeque ribs and sweet ice tea.

www.ingramcontent.com/pod-product-compliance
Lightning Source LLC
Chambersburg PA
CBHW042032220326
41599CB00044BA/7235